ACCESSIBLE AND USABLE BUILDINGS AND FACILITIES

ICC A117.1-2009

ANSI American National Standard

International Code Council
500 New Jersey Avenue, NW, 6th Floor
Washington, DC 20001

Approved October 20, 2010

American National Standard Institute
25 West 43rd Street
New York, NY 10036

Accessible and Usable Buildings and Facilities
(ICC A117.1-2009)

First Printing: January 2011

ISBN: 978-158001-918-7

COPYRIGHT ® 2010
By
INTERNATIONAL CODE COUNCIL, INC.

PRINTED IN THE U.S.A.

AMERICAN NATIONAL STANDARD

Approval of an American National Standard requires verification by ANSI that the requirements for due process, consensus, and other criteria for approval have been met by the standards developer.

Consensus is established when, in the judgement of the ANSI Board of Standards Review, substantial agreement has been reached by directly and materially affected interests. Substantial agreement means much more than a simple majority, but not necessarily unanimity. Consensus requires that all views and objections be considered, and that a concerted effort be made toward their resolution.

The use of American National Standards is completely voluntary; their existence does not in any respect preclude anyone, whether he or she has approved the standards or not, from manufacturing, marketing, purchasing, or using products, processes, or procedures not conforming to the standards.

The American National Standards Institute does not develop standards and will in no circumstances give an interpretation of any American National Standard. Moreover, no person shall have the right or authority to issue an interpretation of an American National Standard in the name of the American National Standards Institute. Requests for interpretations should be addressed to the secretariat or sponsor whose name appears on the title page of this standard.

CAUTION NOTICE: This American National Standard may be revised or withdrawn at any time. The procedures of the American National Standards Institute require that action be taken periodically to reaffirm, revise, or withdraw this standard. Purchasers of American National Standards may receive current information on all standards by calling or writing the American National Standards Institute.

FOREWORD

(The information contained in this foreword is not part of this American National Standard (ANS) and has not been processed in accordance with ANSI's requirements for an ANS. As such, this foreword may contain material that has not been subjected to public review or a consensus process. In addition, it does not contain requirements necessary for conformance to the standard.)

Development

The 1961 edition of ANSI Standard A117.1 presented the first criteria for accessibility to be approved as an American National Standard and was the result of research conducted by the University of Illinois under a grant from the Easter Seal Research Foundation. The National Easter Seal Society and the President's Committee on Employment of People with Disabilities became members of the Secretariat, and the 1961 edition was reaffirmed in 1971.

In 1974, the U.S. Department of Housing and Urban Development joined the Secretariat and sponsored needed research, which resulted in the 1980 edition. After further revision that included a special effort to remove application criteria (scoping requirements), the 1986 edition was published and, when requested in 1987, the Council of American Building Officials (CABO) assumed the Secretariat. Central to the intent of the change in the Secretariat was the development of a standard that, when adopted as part of a building code, would be compatible with the building code and its enforcement. The 1998 edition largely achieved that goal. The 2009 edition of the standard is the latest example of the A117.1 committee's effort to continue developing a standard that is compatible with the building code. (When CABO was consolidated into the International Code Council (ICC) in 1998, the Secretariat duties were assumed by ICC.)

2009 Edition

New to the 2009 edition are coordinated criteria for the various types of dwelling units that provide a step-down between the unit types; technical requirements for Type C (Visitable) Units; Variable Message Signs (i.e., signs that change the information they show such as gate information in train stations and airports); better consistency of sign requirements regarding when raised characters and braille are required; location of toilet paper dispenser (more design options, recessed fixtures addressed, single point of measurement, etc.); a new chapter for a variety of types of recreational facilities; an index and margin markings that will help users find requirements and identify changes from the 2003 edition. In addition, the new standard continued to provide a level of coordination between the accessible provisions of this standard and the Fair Housing Accessibility Guidelines (FHAG) and the newly released Americans with Disabilities Act and Architectural Barriers Act Accessibility Guidelines (ADA & ABA AG).

ANSI Approval

This Standard was processed and approved for submittal to ANSI by the Accredited Standards Committee A117 on Architectural Features and Site Design of Public Buildings and Residential Structures for Persons with Disabilities. ANSI approved the 2009 edition on October 20, 2010. Committee approval of the Standard does not necessarily imply that all Committee members voted for its approval.

Adoption

ICC A117.1–2009 is available for adoption and use by jurisdictions internationally. Its use within a governmental jurisdiction is intended to be accomplished through adoption by reference in accordance with proceedings establishing the jurisdiction's laws.

Formal Interpretations

Requests for Formal Interpretations on the provisions of ICC A117.1–2009 should be addressed to: ICC, Chicago District Office, 4051 W. Flossmoor Road, Country Club Hills, IL 60478–5795.

Maintenance—Submittal of Proposals

All ICC standards are revised as required by ANSI. Proposals for revising this edition are welcome. Please visit the ICC website at www.iccsafe.org for the official "Call for proposals" announcement. A proposal form and instructions can also be downloaded from www.iccsafe.org.

ICC, its members and those participating in the development of ICC A117.1-2009 do not accept any liability resulting from compliance or noncompliance with the provisions of ICC A117.1-2009. ICC does not have the power or authority to police or enforce compliance with the contents of this standard. Only the governmental body that enacts this standard into law has such authority.

Marginal Markings

Solid vertical lines in the margins within the body of the code indicate a technical change from the requirements of the 2003 edition. Deletion indicators in the form of an arrow (➡) are provided in the margin where an entire section, paragraph, exception or table has been deleted or an item in a list of items or a table has been deleted.

Accredited Standards Committee A117 on Architectural Features and Site Design of Public Buildings and Residential Structures for Persons with Disabilities

At the time of ANSI approval, the A117.1 Committee consisted of the following members:

Chair. . . . . Kenneth M. Schoonover, PE
Vice Chair . . . . Vacant
A117 Committee Secretary . . . . Jay Woodward

Organizational Member	Representative
Accessibility Equipment Manufacturers Association (AEMA) (**PD**)	Kevin Brinkman Robert Murphy (Alt)
American Bankers Association (ABA) (**BO**)	Virginia E. O'Neil Nessa Feddis (Alt)
American Council of the Blind (ACB) (**CU**)	Patricia Beattie Eric Bridges (Alt)
American Hotel and Lodging Association (AHLA) (**BO**)	Gerald Gross, AIA, FARA Kevin Maher (Alt)
American Institute of Architects (AIA) (**P**)	David C. Collins, FAIA Larry M. Schneider, AIA (Alt)
American Occupational Therapy Association (AOTA) (**P**)	S. Shoshana Shamberg
American Society of Interior Designers (ASID) (**P**).	Samantha McAskill, ASID Barbara J. Huelat, ASID, IIDA (Alt)
American Society of Plumbing Engineers (ASPE) (**P**)	Robert H. Evans, Jr., CIPE/CPD Julius A. Ballanco, P.E. (Alt)
American Society of Safety Engineers (ASSE) (**P**)	Dr. William Marletta John B. Schroering, P.E. C.S.P. (Alt) Mary Winkler, CSP (Alt)
American Society of Theatre Consultants (ASTC) (**P**).	Scott Crossfield, ASTC William Conner, ASTC (Alt) R. Duane Wilson, ASTC (Alt)
Association for Education & Rehabilitation of the Blind & Visually Impaired (AERBVI) (**P**)	Billie Louise "Beezy" Bentzen, PhD Helen Elias (Alt)
Brain Injury Association of America (BIAA) (**CU**) . . .	Robert Dale Lynch, FAIA Greg Ayotte (Alt)
Builders Hardware Manufacturers Association, Inc. (BHMA) (**PD**)	Michael Tierney Richard Hudnut (Alt)
Building Owners and Managers Association International (BOMA) (**BO**)	Lawrence G. Perry, AIA Ron Burton (Alt)
Disability Rights Education and Defense Fund (DREDF) (**CU**)	Marilyn Golden Logan Hopper (Alt)
Hearing Loss Association of America (HLAA) (**CU**)	Sharon Toji Brenda Battat (Alt)
International Association of Amusement Parks and Attractions (IAAPA) (**BO**)	John Paul Scott, AIA, NCARB Stephanine See (Alt)
International Code Council (ICC) (**R**)	Kimberly Paarlberg RA Phil Hahn, (Alt)

International Sign Association (ISA) **(PD)** . . . . Teresa Cox
Bill Dundas (Alt)
Mike Santos (Alt)
John Souter, PhD. (Alt)

Little People of America, Inc. (LPA) **(CU)** . . . . Tricia Mason

Montgomery County Department of Permitting
Services (MCDPS) **(R)** . . . . Thomas Heiderer

National Association of
Home Builders (NAHB) **(BO)** . . . . Steve Orlowski
Larry Brown (Alt)
Don Surrena, CBO (Alt)

National Association of the Deaf (NAD) **(CU)** . . . . . . Neil McDevitt
Rosaline Crawford (Alt)

National Conference of States on
Building Codes and Standards (NCSBCS) **(R)** . . . Curt Wiehle

National Electrical Manufacturers
Association (NEMA) **(PD)** . . . . Rodger Reiswig, SET
Jack McNamara (Alt)

National Elevator Industry, Inc. (NEII) **(PD)** . . . . Brian D. Black
Barry Blackaby (Alt)
George A. Kappenhagen (Alt)

National Fire Protection Association (NFPA) **(R)** . . . Allan B. Fraser
Ron Coté, PE (Alt)

National Multi Housing Council (NMHC) **(BO)** . . . . . Ronald G. Nickson

New Mexico Governor's Commission on
Disability (NMGCD) **(CU)** . . . . Hope Reed
Anthony H. Alarid (Alt)

Paralyzed Veterans of America (PVA) **(CU)** . . . . Mark H. Lichter, AIA (Alt)
Frank Menendez (Alt)

Plumbing Manufacturers Institute (PMI) **(PD)** . . . . . . Charles Hernandez
David Hagopian (Alt)

Society for Environmental Graphic
Design (SEGD) **(P)** . . . . Kenneth A. Ethridge, AIA, RIBA
Craig Berger (Alt)
Ann Makowski (Alt)
Dave Miller (Alt)

Stairway Manufacturers Association (SMA) **(PD)** . . . David Cooper
Paul Wishnoff (Alt)

United Cerebral Palsy
Association, Inc. (UCPA) **(CU)** . . . . Gina Hilberry
Maureen Fitzgerald (Alt)
Janna Starr (Alt)

United Spinal Association **(CU)** . . . . Dominic Marinelli
John Rooney (Alt)

U.S. Architectural & Transportation Barriers
Compliance (Access) Board (ATBCB) **(R)** . . . . Marsha K. Mazz
Jim Pecht (Alt)

U.S. Department of Agriculture (USDA) **(R)** . . . . William Downs
Meghan Walsh (Alt)

U.S. Department of Housing and
Urban Development (HUD) **(R)** . . . . Cheryl D. Kent
Louis F. Borray (Alt)

World Institute on Disability (WID) **(CU)** . . . . Hale Zukas

Individual Members

Shahriar Amiri CBO (**P**)

Todd Andersen AIA (**P**)

George P. McAllister, Jr. (**P**)

Jake L. Pauls, CPE (**P**)

Ed Roether (**P**)

John P. S. Salmen, AIA (**P**)

Kenneth M. Schoonover, P.E. (**P**)

Acknowledgment

The updating of this standard over the past 6 years could only be accomplished by the hard work of not only the current committee members listed at the time of approval but also the many committee members who participated and contributed to the process over the course of development. ICC recognizes their contributions as well as those of the participants who, although not on the committee, provided valuable input during this update cycle.

INTEREST CATEGORIES

Builder/Owner/Operator (BO) – Members in this category include those in the private sector involved in the development, construction, ownership and operation of buildings or facilities; and their respective associations.

Consumer/User (CU) – Members in this category include those with disabilities, or others who require accessibility features in the built environment for access to buildings, facilities and sites; and their respective associations.

Producer/Distributor (PD) – Members in this category include those involved in manufacturing, distributing, or sales of products; and their respective associations.

Professional (P) – Members in this category include those qualified to engage in the development of the body of knowledge and policy relevant to their area of practice, such as research, testing, consulting, education, engineering or design; and their respective associations.

Regulatory (R) – Members in this category include federal agencies, representatives of regulatory agencies or organizations that promulgate or enforce codes or standards; and their respective associations.

Individual Expert (IE) (Nonvoting) – Members in this category are individual experts selected to assist the consensus body. Individual experts shall serve for a renewable term of one year and shall be subject to approval by vote of the consensus body. Individual experts shall have no vote.

Category	Number
Builder/Owner/Operator – **(BO)**	6
Consumer/User – **(CU)**	11
Professional – **(P)**	15
Producer/Distributor – **(PD)**	7
Regulatory – **(R)**	7
TOTAL	**46**

Contents

Chapter 1. Application and Administration 1

101 Purpose. 1

102 Anthropometric Provisions 1

103 Compliance Alternatives 1

104 Conventions 1

105 Referenced Documents 1

106 Definitions 3

Chapter 2. Scoping 5

201 General 5

202 Dwelling and Sleeping Units 5

203 Administration 5

Chapter 3. Building Blocks 7

301 General 7

302 Floor Surfaces. 7

303 Changes in Level 7

304 Turning Space. 7

305 Clear Floor Space 8

306 Knee and Toe Clearance 9

307 Protruding Objects 9

308 Reach Ranges 11

309 Operable Parts 14

Chapter 4. Accessible Routes. 15

401 General 15

402 Accessible Routes. 15

403 Walking Surfaces 15

404 Doors and Doorways. 15

405 Ramps. 22

406 Curb Ramps 24

407 Elevators. 26

408 Limited-use/Limited-application Elevators. 32

409 Private Residence Elevators 34

410 Platform Lifts 37

Chapter 5. General Site and Building Elements 39

501 General 39

502 Parking Spaces. 39

503 Passenger Loading Zones 40

504 Stairways. 40

505 Handrails 41

506 Windows 44

Chapter 6. Plumbing Elements and Facilities 45

601 General 45

602 Drinking Fountains 45

603 Toilet and Bathing Rooms 46

604 Water Closets and Toilet Compartments 46

605 Urinals 53

606 Lavatories and Sinks 53

607 Bathtubs 54

608 Shower Compartments 56

609 Grab Bars 60

610 Seats 61

611 Washing Machines and Clothes Dryers 62

612 Saunas and Steam Rooms 62

Chapter 7. Communication Elements and Features 65

701 General 65

702 Alarms 65

703 Signs 65

704 Telephones 72

705 Detectable Warnings 73

706 Assistive Listening Systems 73

707 Automatic Teller Machines (ATMs) and Fare Machines 74

708 Two-way Communication Systems 75

Chapter 8. Special Rooms and Spaces 77

801 General 77

802 Assembly Areas 77

803 Dressing, Fitting, and Locker Rooms 81

804 Kitchens and Kitchenettes 81

805 Transportation Facilities 83

806 Holding Cells and Housing Cells 85

807 Courtrooms 85

Chapter 9. Built-In Furnishings and Equipment 87

901 General 87

902 Dining Surfaces and Work Surfaces 87

903 Benches 87

904 Sales and Service Counters 87

905 Storage Facilities 89

Chapter 10. Dwelling Units and Sleeping Units 91

1001 General 91

1002 Accessible Units 91

1003 Type A Units 92

1004 Type B Units 98

1005 Type C (Visitable) Units 105

1006 Units with Accessible Communication Features 106

Chapter 11. Recreational Facilities **107**

1101 General 107

1102 Amusement Rides 107

1103 Recreational Boating Facilities 108

1104 Exercise Machines and Equipment 111

1105 Fishing Piers and Platforms 111

1106 Golf Facilities 113

1107 Miniature Golf Facilities 113

1108 Play Areas 113

1109 Swimming Pools, Wading Pools, Hot Tubs and Spas 117

1110 Shooting Facilities with Firing Positions 122

Index **123**

List of Figures

Chapter 1. Application and Administration 1

Figure 104.3   Graphic Convention for Figures 2

Chapter 2. Scoping (No figures) 5

Chapter 3. Building Blocks 7

Figure 302.2   Carpet on Floor Surfaces 7

Figure 302.3   Openings in Floor Surfaces. 7

Figure 303.2   Carpet on Floor Surfaces 7

Figure 303.3   Beveled Changes in Level 7

Figure 304.3   Size of Turning Space. 8

Figure 305.3   Size of Clear Floor Space 8

Figure 305.5   Position of Clear Floor Space 9

Figure 305.7   Maneuvering Clearance in an Alcove .9

Figure 306.2   Toe Clearance. 10

Figure 306.3   Knee Clearance 10

Figure 307.2   Limits of Protruding Objects 11

Figure 307.3   Post-mounted Protruding Objects. 12

Figure 307.4   Reduced Vertical Clearance 11

Figure 308.2.1   Unobstructed Forward Reach 11

Figure 308.2.2   Obstructed High Forward Reach. 13

Figure 308.3.1   Unobstructed Side Reach 13

Figure 308.3.2   Obstructed High Side Reach. 13

Chapter 4. Accessible Routes 15

Figure 403.5   Clear Width of an Accessible Route .15

Figure 403.5.1   Clear Width at 180° Turn. 16

Figure 404.2.2   Clear Width of Doorways 16

Table 404.2.3.2   Maneuvering Clearances at Manual Swinging Doors 17

Figure 404.2.3.2   Maneuvering Clearances at Manual Swinging Doors 18

Table 404.2.3.3   Manuevering Clearances at Sliding and Folding Doors 17

Figure 404.2.3.3   Maneuvering Clearance at Sliding and Folding Doors 17

Table 404.2.3.4   Maneuvering Clearances for Doorways without Doors. 19

Figure 404.2.3.4   Maneuvering Clearance at Doorways without Doors 19

Figure 404.2.3.5   Maneuvering Clearance at Recessed Doors 20

Figure 404.2.5   Two Doors in a Series. 21

Table 405.2   Allowable Ramp Dimensions for Construction in Existing Sites, Buildings and Facilities 22

Figure 405.7   Ramp Landings. 22

Figure 405.9   Edge Protection—Limited Drop Off. .23

Figure 405.9.1   Extended Floor Surface 23

Figure 405.9.2   Ramp Edge Protection. 24

Figure 406.2   Counter Slope of Surfaces Adjacent to Curb Ramps. 24

Figure 406.3   Sides of Curb Ramps. 24

Figure 406.7   Landings. 25

Figure 406.10   Diagonal Curb Ramps 25

Figure 406.11   Islands 26

Figure 407.2.1.1   Height of Elevator Call Buttons 27

Figure 407.2.1.7   Destination-oriented Elevator Indication 27

Figure 407.2.2.2   Elevator Visible Signals 28

Figure 407.2.3.1   Floor Designation. 29

Figure 407.2.3.2   Destination-oriented Elevator Car Identification. 29

Table 407.4.1   Minimum Dimensions of Elevator Cars 30

Figure 407.4.1   Inside Dimensions of Elevator Cars. 31

Figure 407.4.6.2   Elevator Car Control Buttons 30

Table 407.4.7.1.3   Control Button Identification. 33

Figure 408.3.3   Door Location for Limited Use/Limited Application
                                  (LULA) Elevators 35

Figure 408.4.1   Inside Dimensions of Limited Use/Limited
                                  Application (LULA) Elevator Cars. 36

Figure 409.4.6.3   Location of Controls in Private Residence Elevators 37

Figure 410.2.1   Platform Lift Doors and Gates 38

Chapter 5. General Site and Building Elements. 39

Figure 502.2   Vehicle Parking Space Size. 39

Figure 502.4   Parking Space Access Aisle 39

Figure 503.3   Passenger Loading Zone Acess Aisle 40

Figure 504.2   Treads and Risers for Accessible Stairways 41

Figure 504.5   Stair Nosings 41

Figure 505.4   Handrail Height 42

Figure 505.5   Handrail Clearance 42

Figure 505.7   Handrail Cross Section 43

Figure 505.10.1   Top and Bottom Handrail Extensions at Ramps 43

Figure 505.10.2   Top Handrail Extensions at Stairs 43

Figure 505.10.3   Bottom Handrail Extensions at Stairs. 44

Chapter 6. Plumbing Elements and Facilities 45

Figure 602.2   Parallel Approach at Drinking Fountains
                                  Primarily for Children's Use (Exception 2) 45

Figure 602.5   Drinking Fountain Spout Location 45

Table 603.6   Maximum Reach Depth and Height. 46

Figure 604.2   Water Closet Location 46

Figure 604.3   Size of Clearance for Water Closet 47

Figure 604.4   Water Closet Seat Height. 47

Figure 604.5.1	Side Wall Grab Bar for Water Closet	47
Figure 604.5.2	Rear Wall Grab Bar for Water Closet	48
Figure 604.7	Dispenser Outlet Location	49
Figure 604.9.2	Wheelchair Accessible Toilet Compartments	49
Table 604.9.3.1	Door Opening Location	50
Figure 604.9.3.1	Wheelchair Accessible Compartment Door Openings	50
Figure 604.9.3.1(C)	Wheelchair Accessible Compartment Door Openings—Alternate	51
Figure 604.9.5	Wheelchair Accessible Compartment Toe Clearance	51
Figure 604.10	Ambulatory Accessible Compartment	52
Figure 604.11.2	Children's Water Closet Location	52
Figure 604.11.4	Children's Water Closet Height	52
Figure 604.11.7	Children's Dispenser Outlet Location	53
Figure 605.2	Height of Urinals	53
Figure 606.3	Height of Lavatories and Sinks	54
Figure 607.2	Clearance for Bathtubs	54
Figure 607.4.1	Grab Bars for Bathtubs with Permanent Seats	55
Figure 607.4.2	Grab Bars for Bathtubs without Permanent Seats	55
Figure 607.5	Location of Bathtub Controls	56
Figure 608.2.1	Transfer-type Shower Compartment Size and Clearance	56
Figure 608.2.2	Standard Roll-in-type Shower Compartment Size and Clearance	57
Figure 608.2.3	Alternate Roll-in-type Shower Compartment Size and Clearance	57
Figure 608.3.1	Grab Bars in Transfer-type Showers	57
Figure 608.3.2	Grab Bars in Standard Roll-in-type Showers	58
Figure 608.3.3	Grab Bars in Alternate Roll-in-type Showers	58
Figure 608.4.1	Transfer-type Shower Controls and Handshower Location	58
Figure 608.4.2	Standard Roll-in-type Shower Control and Handshower Location	58
Figure 608.4.3	Alternate Roll-in-type Shower Control and Handshower Location	59
Figure 609.2	Size of Grab Bars	60
Figure 609.3	Spacing of Grab Bars	60
Figure 609.4.2	Position of Children's Grab Bars	61
Figure 610.2	Bathtub Seats	61
Figure 610.3.1	Rectangular Shower Compartment Seat	62
Figure 610.3.2	L-shaped Shower Compartment Seat	62
Figure 611.2	Clear Floor Space	63
Figure 611.4	Height of Laundry Equipment	63

Chapter 7. Communication Elements and Features **65**

Table 703.2.4   Visual Character Height 65

Figure 703.3.5   Character Height 66

Figure 703.3.10   Height of Raised Characters above Floor 67

Figure 703.3.11   Location of Signs at Doors 67

Figure 703.4.3   Braille Measurement 68

Table 703.4.3   Braille Dimensions 68

Figure 703.4.4   Position of Braille 69

Figure 703.4.5   Height of Braille Characters Above Floor 69

Figure 703.5   Pictogram Field 69

Figure 703.6.3.1   International Symbol of Accessibility 70

Figure 703.6.3.2   International TTY Symbol 70

Figure 703.6.3.3   International Symbol of Access for Hearing Loss 70

Figure 703.6.3.4   Volume-controlled Telephone 70

Table 703.7.4   Low Resolution VMS Character Height 71

Table 703.7.5   Pixel Count for Low Resolution VMS Signage 71

Figure 703.7.5   Low Resolution VMS Signage Character 72

Figure 704.2.1   Clear Floor Space for Telephones 73

Figure 705.5   Truncated Dome Size and Spacing 74

Figure 707.5   Numeric Key Layout 74

Table 707.6.1   Raised Symbols 75

Chapter 8. Special Rooms and Spaces **77**

Figure 802.3   Width of a Wheelchair Space in Assembly Areas 77

Figure 802.4   Depth of a Wheelchair Space in Assembly Areas 77

Figure 802.9.1.1   Lines of Sight over the Heads of Seated Spectators 78

Figure 802.9.1.2   Lines of Sight between the Heads of Seated Spectators 79

Figure 802.9.2   Line of Sight Over Standing Spectators 79

Table 802.9.2.2   Required Wheelchair Space Location Elevation
                                    Over Standing Spectators 80

Table 802.10   Wheelchair Space Location Dispersion 80

Figure 804.2.1   Pass-through Kitchen Clearance 81

Figure 804.2.2   U-shaped Kitchen Clearance 82

Figure 805.2.2   Size of Bus Boarding and Alighting Areas 83

Figure 805.3   Bus Shelters 84

Figure 805.10   Track Crossings 84

Chapter 9. Built-In Furnishings and Equipment **87**

Figure 903   Benches 88

Figure 904.4.2   Height of Checkout Counters 88

Chapter 10. Dwelling Units and Sleeping Units **91**

Figure 1003.11.2.4   Water Closets in Type A Units 94

Figure 1003.11.2.5.1   Clearance for Bathtubs in Type A Units 95

Figure 1003.11.2.5.2   Standard Roll-in-type Shower Compartment in Type A Units 95

Figure 1003.12.1.1   Minimum Kitchen Clearance in Type A Units 95

Figure 1003.12.1.2   U-shaped Kitchen Clearance in Type A Units. 97

Figure 1003.12.3   Work Surface in Kitchen for Type A Units. 97

Figure 1003.12.4   Kitchen Sink for Type A Units 97

Figure 1004.11.1.1   Swing-up Grab Bar for Water Closet. 100

Figure 1004.11.3.1.1   Lavatory in Type B Units – Option A Bathrooms. 101

Figure 1004.11.3.1.2   Clearance at Water Closets in Type B Units. 102

Figure 1004.11.3.1.3.1   Parallel Approach Bathtub in Type B Units– Option A Bathrooms 102

Figure 1004.11.3.1.3.2   Forward Approach Bathtub in Type B Units– Option A Bathrooms 103

Figure 1004.11.3.1.3.3   Transfer-type Shower Compartment in Type B Units 103

Figure 1004.11.3.2.1   Lavatory in Type B Units – Option B Bathrooms. 103

Figure 1004.11.3.2.3.1   Bathroom Clearance in Type B Units – Option B Bathrooms . . .103

Figure 1004.12.1.1   Minimum Kitchen Clearance in Type B Units 104

Figure 1004.12.1.2   U-shaped Kitchen Clearance in Type B Units. 104

Chapter 11. Recreational Facilities. **107**

Figure 1102.4.4.3   Protrusions in Wheelchair Spaces in Amusement Rides 108

Figure 1103.3.1(A)   Boat Slip Clearance 109

Figure 1103.3.1(B)   (Exception 1) Clear Pier Space Reduction at Boat Slips 109

Figure 1103.3.1(C)   (Exception 2) Edge Protection at Boat Slips 109

Figure 1103.3.2(A)   Boarding Pier Clearance. 111

Figure 1103.3.2(B)   (Exception 1) Clear Pier Space Reduction at Boarding Piers . . .111

Figure 1103.3.2(C)   (Exception 2) Edge Protection at Boarding Piers 112

Figure 1105.3.2   Extended Ground or Deck Surface at Fishing Piers and Platforms. 112

Figure 1107.3.2   Golf Club Reach Range Area 114

Table 1108.3.2.1.2   Number and Types of Ground Level Play Components Required to be on Accessible Routes 115

Figure 1108.4.2.1   Transfer Platforms. 116

Figure 1108.4.2.2   Transfer Steps. 117

Figure 1109.2.2   Pool Lift Seat Location 118

Figure 1109.2.3   Clear Deck Space at Pool Lifts 118

Figure 1109.2.4   Pool Lift Seat Height 118

Figure 1109.2.8   Pool Lift Submerged Depth. 119

Figure 1109.3.2   Sloped Entry Submerged Depth. 119

Figure 1109.3.3   Handrails for Sloped Entry 119

Figure 1109.4.1   Clear Deck Space at Transfer Walls. 120

Figure 1109.4.2   Transfer Wall Height 120

Figure 1109.4.3   Depth and Length of Transfer Walls 120

Figure 1109.4.5   Grab Bars for Transfer Walls 121

Figure 1109.5.1   Size of Transfer Platform 121

Figure 1109.5.2   Clear Deck Space at Transfer Platform 121

Figure 1109.5.4   Transfer Steps 121

Figure 1109.5.6   Size of Transfer Steps 122

Figure 110.5.7   Grab Bars 122

Chapter 1. Application and Administration

101 Purpose

The technical criteria in Chapters 3 through 9, Sections 1002, 1003 and 1006 and Chapter 11 of this standard make sites, facilities, buildings and elements accessible to and usable by people with such physical disabilities as the inability to walk, difficulty walking, reliance on walking aids, blindness and visual impairment, deafness and hearing impairment, incoordination, reaching and manipulation disabilities, lack of stamina, difficulty interpreting and reacting to sensory information, and extremes of physical size. The intent of these sections is to allow a person with a physical disability to independently get to, enter, and use a site, facility, building, or element.

Section 1004 of this standard provides criteria for Type B units. These criteria are intended to be consistent with the intent of the criteria of the U.S. Department of Housing and Urban Development (HUD) Fair Housing Accessibility Guidelines. The Type B units are intended to supplement, not replace, Accessible units or Type A units as specified in this standard.

Section 1005 of this standard provides criteria for minimal accessibility features for one and two family dwelling units and townhouses which are not covered by the U.S. Department of Housing and Urban Development (HUD) Fair Housing Accessibility Guidelines.

This standard is intended for adoption by government agencies and by organizations setting model codes to achieve uniformity in the technical design criteria in building codes and other regulations.

101.1 Applicability. Sites, facilities, buildings, and elements required to be accessible shall comply with the applicable provisions of Chapters 3 through 9 and Chapter 11. Dwelling units and sleeping units shall comply with the applicable provisions of Chapter 10.

102 Anthropometric Provisions

The technical criteria in this standard are based on adult dimensions and anthropometrics. This standard also contains technical criteria based on children's dimensions and anthropometrics for drinking fountains, water closets, toilet compartments, lavatories and sinks, dining surfaces, work surfaces and benches.

103 Compliance Alternatives

Nothing in this standard is intended to prevent the use of designs, products, or technologies as alternatives to those prescribed by this standard, provided they result in equivalent or greater accessibility and such equivalency is approved by the administrative authority adopting this standard.

104 Conventions

104.1 General. Where specific criteria of this standard differ from the general criteria of this standard, the specific criteria shall apply.

104.2 Dimensions. Dimensions that are not stated as "maximum" or "minimum" are absolute. All dimensions are subject to conventional industry tolerances.

104.3 Figures. Unless specifically stated, figures included herein are provided for informational purposes only and are not considered part of the standard.

104.4 Floor or Floor Surface. The terms floor or floor surface refer to the finish floor surface or ground surface, as applicable.

104.5 Referenced Sections. Unless specifically stated otherwise, a reference to another section or subsection within this standard includes all subsections of the referenced section or subsection.

105 Referenced Documents

105.1 General. The documents listed in Section 105.2 shall be considered part of this standard to the prescribed extent of each such reference. Where criteria in this standard differ from those of these referenced documents, the criteria of this standard shall apply.

105.2 Documents.

105.2.1 Manual on Uniform Traffic Control Devices: MUTCD-2003 (The Federal Highway Administration, Office of Transportation Operations, Room 3408, 400 7th Street, S.W., Washington, DC 20590).

105.2.2 National Fire Alarm Code: NFPA 72-2007 (National Fire Protection Association, 1 Batterymarch Park, Quincy, MA 02269-9101).

105.2.3 Power Assist and Low Energy Power Operated Doors: ANSI/BHMA A156.19-2007. (Builders Hardware Manufacturers' Association, 355 Lexington Avenue, 15th Floor, New York, NY 10017).

105.2.4 Power Operated Pedestrian Doors: ANSI/BHMA A156.10-2005 (Builders Hardware Manufacturers' Association, 355 Lexington Avenue, 15th Floor, New York, NY 10017).

105.2.5 Safety Code for Elevators and Escalators: ASME A17.1-2007/CSA B44-07 (American Society of Mechanical Engineers International, Three Park Avenue, New York, NY 10016-5990).

Convention	Description
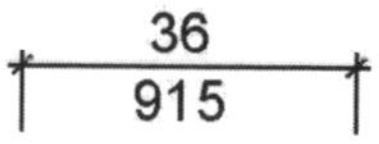 $\frac{36}{915}$	dimension showing English units (in inches unless otherwise specified) above the line and SI units (in millimeters unless otherwise specified) below the line
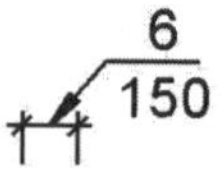 $\frac{6}{150}$	dimension for small measurements
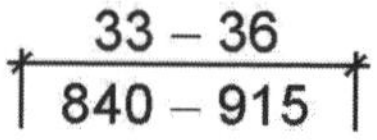 $\frac{33-36}{840-915}$	dimension showing a range with minimum – maximum
min	minimum
max	maximum
>	greater than
≥	greater than or equal to
<	less than
≤	less than or equal to
 - - - - - - - -	boundary of clear floor space or maneuvering clearance
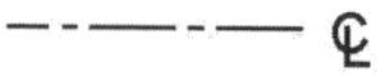 - - — - - — ₵	centerline
 - - — - - - —	a permitted element or its extension
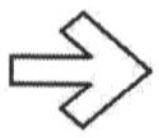 ⇨	direction of travel or approach
▬▬▬▬	a wall, floor, ceiling or other element cut in section or plan
�	a highlighted element in elevation or plan
 ▨▨▨	location zone of element, control or feature

FIG. 104.3
GRAPHIC CONVENTION FOR FIGURES

105.2.6 Safety Standard for Platform Lifts and Stairway Chairlifts: ASME A18.1-2005 (American Society of Mechanical Engineers International, Three Park Avenue, New York, NY 10016-5990).

105.2.7 Performance Criteria for Accessible Communications Entry Systems. ANSI/DASMA 303-2006. (Door and Access Systems Manufacturers Association, 1300 Sumner Avenue, Cleveland, OH 44115-2851).

105.2.8 Standard Specification for Impact Attenuation of Surface Systems Under and Around Playground Equipment ASTM F 1292-99. (ASTM International, 100 Barr Harbor Drive, PO Box C700, West Conshohocken, PA, 19428-2959).

105.2.9 Standard Specification for Impact Attenuation of Surfacing Materials Within the Use Zone of Playground Equipment ASTM F 1292-04 (ASTM International, 100 Barr Harbor Drive, PO Box C700, West Conshohocken, PA, 19428-2959).

105.2.10 Standard Consumer Safety Performance Specification for Playground Equipment for Public Use ASTM F 1487-01 (ASTM International, 100 Barr Harbor Drive, PO Box C700, West Conshohocken, PA, 19428-2959).

105.2.11 Americans with Disabilities Act (ADA) Accessibility Guidelines for Transportation Vehicles 36 CFR 1192 published in 56 Federal Register 45558, September 6, 1991 (United States Access Board, 1331 F Street, NW, Suite 1000, Washington, DC 20004-1111).

106 Definitions

106.1 General. For the purpose of this standard, the terms listed in Section 106.5 have the indicated meaning.

106.2 Terms Defined in Referenced Documents. Terms specifically defined in a referenced document, and not defined in this section, shall have the specified meaning from the referenced document.

106.3 Undefined Terms. The meaning of terms not specifically defined in this standard or in a referenced document shall be as defined by collegiate dictionaries in the sense that the context implies.

106.4 Interchangeability. Words, terms, and phrases used in the singular include the plural, and those used in the plural include the singular.

106.5 Defined Terms.

accessible: Describes a site, building, facility, or portion thereof that complies with this standard.

administrative authority: A jurisdictional body that adopts or enforces regulations and standards for the design, construction, or operation of buildings and facilities.

amusement attraction: Any facility, or portion of a facility, located within an amusement park or theme park which provides amusement without the use of an amusement device. Amusement attractions include, but are not limited to, fun houses, barrels, and other attractions without seats.

amusement ride: A system that moves persons through a fixed course within a defined area for the purpose of amusement.

amusement ride seat: A seat that is built-in or mechanically fastened to an amusement ride intended to be occupied by one or more passengers.

area of sport activity: That portion of a room or space where the play or practice of a sport occurs.

boarding pier: A portion of a pier where a boat is temporarily secured for the purpose of embarking or disembarking.

boat launch ramp: A sloped surface designed for launching and retrieving trailered boats and other water craft to and from a body of water.

boat slip: That portion of a pier, main pier, finger pier, or float where a boat is moored for the purpose of berthing, embarking, or disembarking.

catch pool: A pool or designated section of a pool used as a terminus for water slide flumes.

characters: Letters, numbers, punctuation marks, and typographic symbols.

children's use: Spaces and elements specifically designed for use primarily by people 12 years old and younger.

circulation path: An exterior or interior way of passage from one place to another for pedestrians.

counter slope: Any slope opposing the running slope of a curb ramp.

cross slope: The slope that is perpendicular to the direction of travel (see running slope).

curb ramp: A short ramp cutting through a curb or built up to it.

destination-oriented elevator system: An elevator system that provides lobby controls for the selection of destination floors, lobby indicators designating which elevator to board, and a car indicator designating the floors at which the car will stop.

detectable warning: A standardized surface feature built in or applied to floor surfaces to warn of hazards on a circulation path.

dwelling unit: A single unit providing complete, independent living facilities for one or more persons including permanent provisions for living, sleeping, eating, cooking and sanitation.

element: An architectural or mechanical component of a building, facility, space, or site.

elevated play component: A play component that is approached above or below grade and that is part of a composite play structure consisting of two or more play components attached or functionally linked to create an integrated unit providing more than one play activity.

elevator car call sequential step scanning: A technology used to enter a car call by means of an up or down floor selection button.

facility: All or any portion of a building, structure, site improvements, elements, and pedestrian routes or vehicular ways located on a site.

gangway: A variable-sloped pedestrian walkway that links a fixed structure or land with a floating structure. Gangways that connect to vessels are not addressed by this document.

golf car passage: A continuous passage on which a motorized golf car can operate.

ground level play component: A play component that is approached and exited at the ground level.

habitable: A space in a building for living, sleeping, eating or cooking. Bathrooms, toilet rooms, closets, halls, storage or utility spaces and similar areas are not considered habitable spaces.

key surface: The surface or plane of any key or button that must be touched to activate or deactivate an operable part or a machine function or enter data.

marked crossing: A crosswalk or other identified path intended for pedestrian use in crossing a vehicular way.

operable part: A component of an element used to insert or withdraw objects, or to activate, deactivate, or adjust the element.

pictogram: A pictorial symbol that represents activities, facilities, or concepts.

play area: A portion of a site containing play components designed and constructed for children.

play component: An element intended to generate specific opportunities for play, socialization, or learning. Play components are manufactured or natural; and are stand-alone or part of a composite play structure.

ramp: A walking surface that has a running slope steeper than 1:20.

running slope: The slope that is parallel to the direction of travel (see cross slope).

sign: An architectural element composed of displayed textual, symbolic, tactile, or pictorial information.

site: A parcel of land bounded by a property line or a designated portion of a public right-of-way.

sleeping unit: A room or space in which people sleep that can also include permanent provisions for living, sleeping, eating, and either sanitation or kitchen facilities but not both. Such rooms and spaces that are also part of a dwelling unit are not sleeping units.

soft contained play structure: A play structure made up of one or more play components where the user enters a fully enclosed play environment that utilizes pliable materials, such as plastic, netting, or fabric.

teeing ground: In golf, the starting place for the hole to be played.

transfer device: Equipment designed to facilitate the transfer of a person from a wheelchair or other mobility aide to and from an amusement ride seat.

TTY: An abbreviation for teletypewriter. Equipment that employs interactive, text-based communications through the transmission of coded signals across the standard telephone network. The term TTY also refers to devices known as text telephones and TDDs.

use zone: The ground level area beneath and immediately adjacent to a play structure or play equipment that is designated by ASTM F 1487 listed in Section 105.2.10, for unrestricted circulation around the play equipment and where it is predicted that a user would land when falling from or exiting the play equipment.

variable message signs (VMS): Electronic signs that have a message with the capacity to change by means of scrolling, streaming, or paging across a background.

variable message sign (VMS) characters: Characters of an electronic sign are composed of pixels in an array. High resolution VMS characters have vertical pixel counts of 16 rows or greater. Low resolution VMS characters have vertical pixel counts of 7 to 15 rows.

vehicular way: A route provided for vehicular traffic.

walk: An exterior pathway with a prepared surface for pedestrian use.

wheelchair space: A space for a single wheelchair and its occupant.

wheelchair space locations: A space for a minimum of a single wheelchair and the associated companion seating. Wheelchair space locations can contain multiple wheelchair spaces and associated companion seating.

Chapter 2. Scoping

201 General

This standard provides technical criteria for making sites, facilities, buildings, and elements accessible. The administrative authority shall provide scoping provisions to specify the extent to which these technical criteria apply. These scoping provisions shall address the application of this standard to: each building and occupancy type; new construction, alterations, temporary facilities, and existing buildings; specific site and building elements; and to multiple elements or spaces provided within a site or building.

202 Dwelling and Sleeping Units

Chapter 10 of this standard contains dwelling unit and sleeping unit criteria for Accessible units, Type A units, Type B units, Type C (Visitable) dwelling units and units with accessible communication features. The administrative authority shall specify, in separate scoping provisions, the extent to which these technical criteria apply. These scoping provisions shall address the types and numbers of units required to comply with each set of unit criteria.

203 Administration

The administrative authority shall provide an appropriate review and approval process to ensure compliance with this standard.

Chapter 3. Building Blocks

301 General

301.1 Scope. The provisions of Chapter 3 shall apply where required by the scoping provisions adopted by the administrative authority or by Chapters 4 through 11.

301.2 Overlap. Unless otherwise specified, clear floor spaces, clearances at fixtures, maneuvering clearances at doors, and turning spaces shall be permitted to overlap.

302 Floor Surfaces

302.1 General. Floor surfaces shall be stable, firm, and slip resistant, and shall comply with Section 302. Changes in level in floor surfaces shall comply with Section 303.

302.2 Carpet. Carpet or carpet tile shall be securely attached and shall have a firm cushion, pad, or backing or no cushion or pad. Carpet or carpet tile shall have a level loop, textured loop, level cut pile, or level cut/uncut pile texture. The pile shall be $^1/_2$ inch (13 mm) maximum in height. Exposed edges of carpet shall be fastened to the floor and shall have trim along the entire length of the exposed edge. Carpet edge trim shall comply with Section 303.

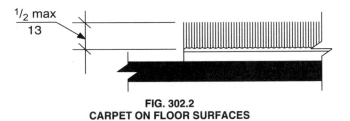

FIG. 302.2
CARPET ON FLOOR SURFACES

302.3 Openings. Openings in floor surfaces shall be of a size that does not permit the passage of a $^1/_2$ inch (13 mm) diameter sphere, except as allowed in Sections 407.4.3, 408.4.3, 409.4.3, 410.4, and 805.10. Elongated openings shall be placed so that the long dimension is perpendicular to the predominant direction of travel.

303 Changes in Level

303.1 General. Changes in level in floor surfaces shall comply with Section 303.

303.2 Vertical. Changes in level of $^1/_4$ inch (6.4 mm) maximum in height shall be permitted to be vertical.

303.3 Beveled. Changes in level greater than $^1/_4$ inch (6.4 mm) in height and not more than $^1/_2$ inch (13 mm) maximum in height shall be beveled with a slope not steeper than 1:2.

303.4 Ramps. Changes in level greater than $^1/_2$ inch (13 mm) in height shall be ramped and shall comply with Section 405 or 406.

Predominant direction of travel

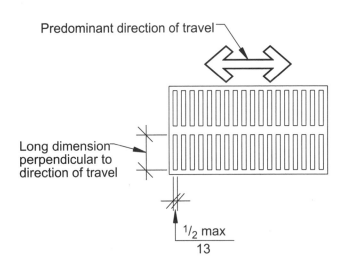

FIG. 302.3
OPENINGS IN FLOOR SURFACES

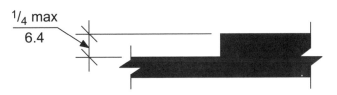

FIG. 303.2
CARPET ON FLOOR SURFACES

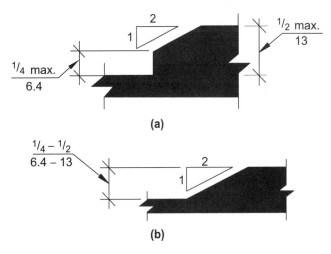

FIG. 303.3
BEVELED CHANGES IN LEVEL

304 Turning Space

304.1 General. A turning space shall comply with Section 304.

304.2 Floor Surface. Floor surfaces of a turning space shall comply with Section 302. Changes in level are not permitted within the turning space.

> **EXCEPTION:** Slopes not steeper than 1:48 shall be permitted.

304.3 Size. Turning spaces shall comply with Section 304.3.1 or 304.3.2.

> **304.3.1 Circular Space.** The turning space shall be a circular space with a 60-inch (1525 mm) minimum diameter. The turning space shall be permitted to include knee and toe clearance complying with Section 306.

> **304.3.2 T-Shaped Space.** The turning space shall be a T-shaped space within a 60-inch (1525 mm) minimum square, with arms and base 36 inches (915 mm) minimum in width. Each arm of the T shall be clear of obstructions 12 inches (305 mm) minimum in each direction, and the base shall be clear of obstructions 24 inches (610 mm) minimum. The turning space shall be permitted to include knee and toe clearance complying with Section 306 only at the end of either the base or one arm.

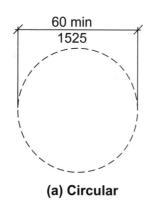

(a) Circular

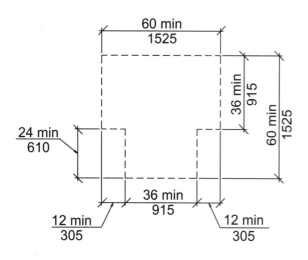

(b) T-shaped

FIG. 304.3
SIZE OF TURNING SPACE

304.4 Door Swing. Unless otherwise specified, doors shall be permitted to swing into turning spaces.

305 Clear Floor Space

305.1 General. A clear floor space shall comply with Section 305.

305.2 Floor Surfaces. Floor surfaces of a clear floor space shall comply with Section 302. Changes in level are not permitted within the clear floor space.

> **EXCEPTION:** Slopes not steeper than 1:48 shall be permitted.

305.3 Size. The clear floor space shall be 48 inches (1220 mm) minimum in length and 30 inches (760 mm) minimum in width.

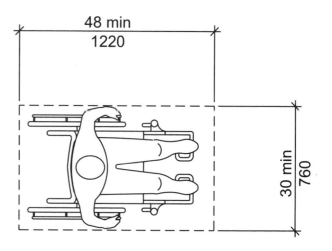

FIG. 305.3
SIZE OF CLEAR FLOOR SPACE

305.4 Knee and Toe Clearance. Unless otherwise specified, clear floor space shall be permitted to include knee and toe clearance complying with Section 306.

305.5 Position. Unless otherwise specified, the clear floor space shall be positioned for either forward or parallel approach to an element.

305.6 Approach. One full, unobstructed side of the clear floor space shall adjoin or overlap an accessible route or adjoin another clear floor space.

305.7 Alcoves. If a clear floor space is in an alcove or otherwise confined on all or part of three sides, additional maneuvering clearances complying with Sections 305.7.1 and 305.7.2 shall be provided, as applicable.

> **305.7.1 Parallel Approach.** Where the clear floor space is positioned for a parallel approach, the alcove shall be 60 inches (1525 mm) minimum in width where the depth exceeds 15 inches (380 mm).

> **305.7.2 Forward Approach.** Where the clear floor space is positioned for a forward approach, the alcove shall be 36 inches (915 mm) minimum in width where the depth exceeds 24 inches (610 mm).

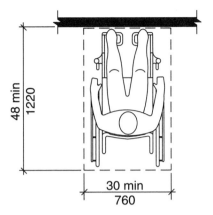

(a) Forward

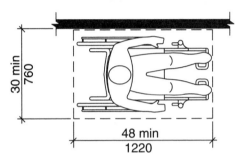

(b) Parallel

FIG. 305.5
POSITION OF CLEAR FLOOR SPACE

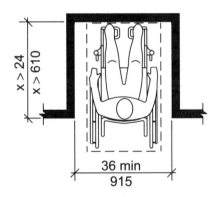

(a) Forward Approach

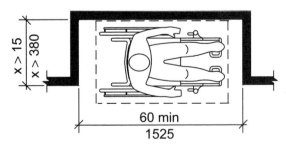

(b) Parallel Approach

FIG. 305.7
MANEUVERING CLEARANCE IN AN ALCOVE

306 Knee and Toe Clearance

306.1 General. Where space beneath an element is included as part of clear floor space at an element, clearance at an element, or a turning space, the space shall comply with Section 306. Additional space shall not be prohibited beneath an element, but shall not be considered as part of the clear floor space or turning space.

306.2 Toe Clearance.

306.2.1 General. Space beneath an element between the floor and 9 inches (230 mm) above the floor shall be considered toe clearance and shall comply with Section 306.2.

306.2.2 Maximum Depth. Toe clearance shall be permitted to extend 25 inches (635 mm) maximum under an element.

306.2.3 Minimum Depth. Where toe clearance is required at an element as part of a clear floor space complying with Section 305, the toe clearance shall extend 17 inches (430 mm) minimum beneath the element.

306.2.4 Additional Clearance. Space extending greater than 6 inches (150 mm) beyond the available knee clearance at 9 inches (230 mm) above the floor shall not be considered toe clearance.

306.2.5 Width. Toe clearance shall be 30 inches (760 mm) minimum in width.

306.3 Knee Clearance.

306.3.1 General. Space beneath an element between 9 inches (230 mm) and 27 inches (685 mm) above the floor shall be considered knee clearance and shall comply with Section 306.3.

306.3.2 Maximum Depth. Knee clearance shall be permitted to extend 25 inches (635 mm) maximum under an element at 9 inches (230 mm) above the floor.

306.3.3 Minimum Depth. Where knee clearance is required beneath an element as part of a clear floor space complying with Section 305, the knee clearance shall be 11 inches (280 mm) minimum in depth at 9 inches (230 mm) above the floor, and 8 inches (205 mm) minimum in depth at 27 inches (685 mm) above the floor.

306.3.4 Clearance Reduction. Between 9 inches (230 mm) and 27 inches (685 mm) above the floor, the knee clearance shall be permitted to be reduced at a rate of 1 inch (25 mm) in depth for each 6 inches (150 mm) in height.

306.3.5 Width. Knee clearance shall be 30 inches (760 mm) minimum in width.

307 Protruding Objects

307.1 General. Protruding objects on circulation paths shall comply with Section 307.

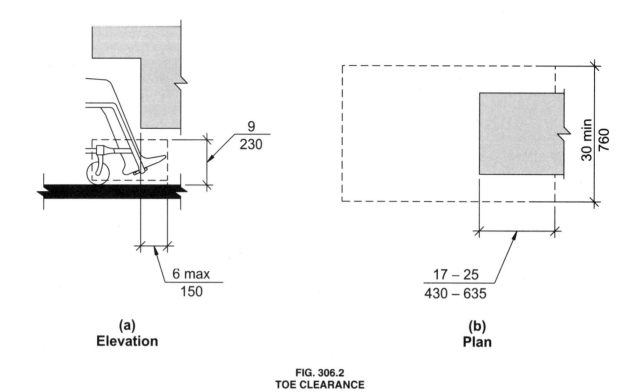

(a)
Elevation

(b)
Plan

FIG. 306.2
TOE CLEARANCE

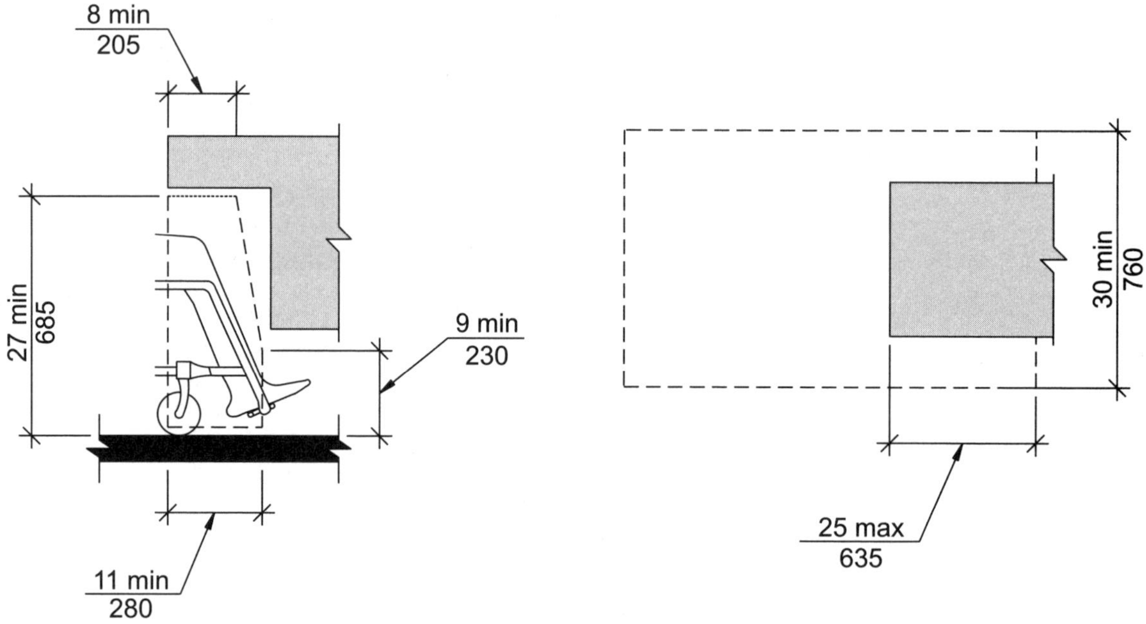

9 / 230

6 max / 150

30 min / 760

17 − 25 / 430 − 635

8 min / 205

27 min / 685

9 min / 230

11 min / 280

30 min / 760

25 max / 635

(a)
Elevation

(b)
Plan

FIG. 306.3
KNEE CLEARANCE

307.2 Protrusion Limits. Objects with leading edges more than 27 inches (685 mm) and not more than 80 inches (2030 mm) above the floor shall protrude 4 inches (100 mm) maximum horizontally into the circulation path.

> **EXCEPTION:** Handrails shall be permitted to protrude 4¹/₂ inches (115 mm) maximum.

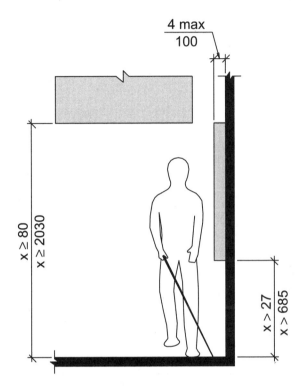

FIG. 307.2
LIMITS OF PROTRUDING OBJECTS

307.3 Post-Mounted Objects. Objects on posts or pylons shall be permitted to overhang 4 inches (100 mm) maximum where more than 27 inches (685 mm) and not more than 80 inches (2030 mm) above the floor. Objects on multiple posts or pylons where the clear distance between the posts or pylons is greater than 12 inches (305 mm) shall have the lowest edge of such object either 27 inches (685 mm) maximum or 80 inches (2030 mm) minimum above the floor.

> **EXCEPTION:** Sloping portions of handrails between the top and bottom riser of stairs and above the ramp run shall not be required to comply with Section 307.3.

307.4 Vertical Clearance. Vertical clearance shall be 80 inches (2030 mm) minimum. Rails or other barriers shall be provided where the vertical clearance is less than 80 inches (2030 mm). The leading edge of such rails or barrier shall be located 27 inches (685 mm) maximum above the floor.

> **EXCEPTION:** Door closers and door stops shall be permitted to be 78 inches (1980 mm) minimum above the floor.

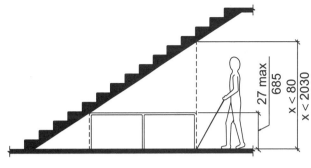

FIG. 307.4
REDUCED VERTICAL CLEARANCE

307.5 Required Clear Width. Protruding objects shall not reduce the clear width required for accessible routes.

308 Reach Ranges

308.1 General. Reach ranges shall comply with Section 308.

308.2 Forward Reach.

308.2.1 Unobstructed. Where a forward reach is unobstructed, the high forward reach shall be 48 inches (1220 mm) maximum and the low forward reach shall be 15 inches (380 mm) minimum above the floor.

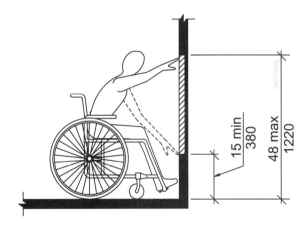

FIG. 308.2.1
UNOBSTRUCTED FORWARD REACH

308.2.2 Obstructed High Reach. Where a high forward reach is over an obstruction, the clear floor space complying with Section 305 shall extend beneath the element for a distance not less than the required reach depth over the obstruction. The high forward reach shall be 48 inches (1220 mm) maximum above the floor where the reach depth is 20 inches (510mm) maximum. Where the reach depth exceeds 20 inches (510 mm), the high forward reach shall be 44 inches (1120 mm) maximum above the floor, and the reach depth shall be 25 inches (635 mm) maximum.

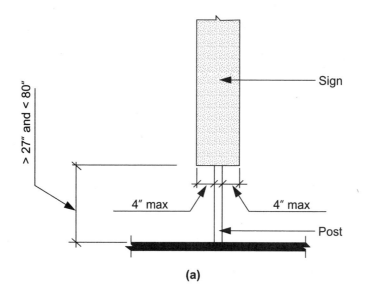

(a)

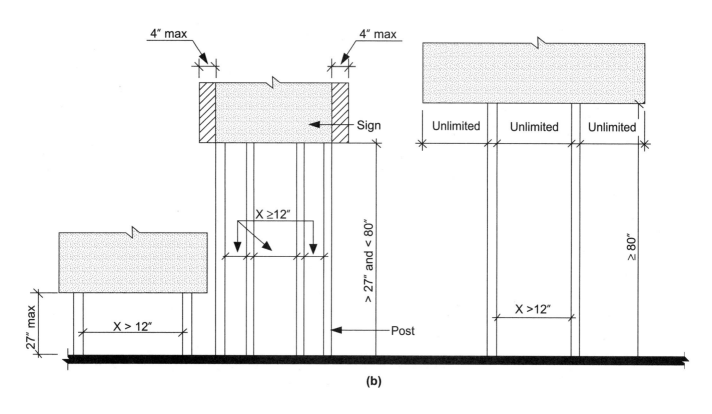

(b)

Fig. 307.3
Post-Mounted Protruding Objects

element, the high side reach shall be 48 inches (1220 mm) maximum and the low side reach shall be 15 inches (380 mm) minimum above the floor.

EXCEPTION: Existing elements that are not altered shall be permitted at 54 inches (1370 mm) maximum above the floor.

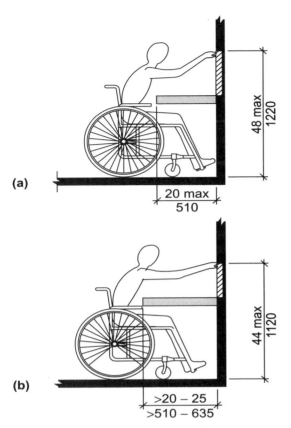

(a)

20 max
510

(b)

>20 – 25
>510 – 635

FIG. 308.2.2
OBSTRUCTED HIGH FORWARD REACH

308.3 Side Reach.

308.3.1 Unobstructed. Where a clear floor space complying with Section 305 allows a parallel approach to an element and the edge of the clear floor space is 10 inches (255 mm) maximum from the

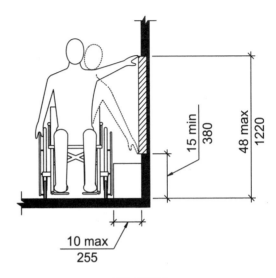

10 max
255

FIG. 308.3.1
UNOBSTRUCTED SIDE REACH

308.3.2 Obstructed High Reach. Where a clear floor space complying with Section 305 allows a parallel approach to an element and the high side reach is over an obstruction, the height of the obstruction shall be 34 inches (865 mm) maximum above the floor and the depth of the obstruction shall be 24 inches (610 mm) maximum. The high side reach shall be 48 inches (1220 mm) maximum above the floor for a reach depth of 10 inches (255 mm) maxi-

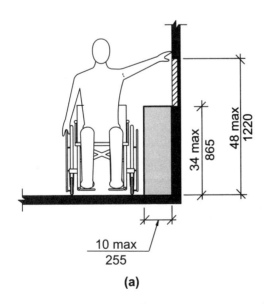

10 max
255

(a)

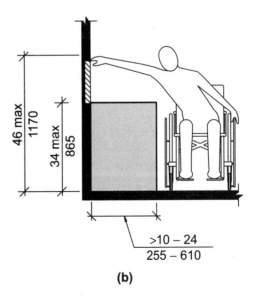

>10 – 24
255 – 610

(b)

FIG. 308.3.2
OBSTRUCTED HIGH SIDE REACH

mum. Where the reach depth exceeds 10 inches (255 mm), the high side reach shall be 46 inches (1170 mm) maximum above the floor for a reach depth of 24 inches (610 mm) maximum.

> **EXCEPTION:** At washing machines and clothes dryers, the height of the obstruction shall be permitted to be 36 inches (915 mm) maximum above the floor.

309 Operable Parts

309.1 General. Operable parts required to be accessible shall comply with Section 309.

309.2 Clear Floor Space. A clear floor space complying with Section 305 shall be provided.

309.3 Height. Operable parts shall be placed within one or more of the reach ranges specified in Section 308.

309.4 Operation. Operable parts shall be operable with one hand and shall not require tight grasping, pinching, or twisting of the wrist. The force required to activate operable parts shall be 5.0 pounds (22.2 N) maximum.

> **EXCEPTION:** Gas pump nozzles shall not be required to provide operable parts that have an activating force of 5.0 pounds (22.2 N) maximum.

Chapter 4. Accessible Routes

401 General

401.1 Scope. Accessible routes required by the scoping provisions adopted by the administrative authority shall comply with the applicable provisions of Chapter 4.

402 Accessible Routes

402.1 General. Accessible routes shall comply with Section 402.

402.2 Components. Accessible routes shall consist of one or more of the following components: Walking surfaces with a slope not steeper than 1:20, doors and doorways, ramps, curb ramps excluding the flared sides, elevators, and platform lifts. All components of an accessible route shall comply with the applicable portions of this standard.

402.3 Revolving Doors, Revolving Gates, and Turnstiles. Revolving doors, revolving gates, and turnstiles shall not be part of an accessible route.

403 Walking Surfaces

403.1 General. Walking surfaces that are a part of an accessible route shall comply with Section 403.

403.2 Floor Surface. Floor surfaces shall comply with Section 302.

403.3 Slope. The running slope of walking surfaces shall not be steeper than 1:20. The cross slope of a walking surface shall not be steeper than 1:48.

403.4 Changes in Level. Changes in level shall comply with Section 303.

403.5 Clear Width. The clear width of an accessible route shall be 36 inches (915 mm) minimum.

> **EXCEPTION:** The clear width shall be permitted to be reduced to 32 inches (815 mm) minimum for a length of 24 inches (610 mm) maximum provided the reduced width segments are separated by segments that are 48 inches (1220 mm) minimum in length and 36 inches (915 mm) minimum in width.

403.5.1 Clear Width at 180 Degree Turn. Where an accessible route makes a 180 degree turn around an object that is less than 48 inches (1220 mm) in width, clear widths shall be 42 inches (1065 mm) minimum approaching the turn, 48 inches (1220 mm) minimum during the turn, and 42 inches (1065 mm) minimum leaving the turn.

> **EXCEPTION:** Section 403.5.1 shall not apply where the clear width during the turn is 60 inches (1525 mm) minimum.

403.5.2 Passing Space. An accessible route with a clear width less than 60 inches (1525 mm) shall provide passing spaces at intervals of 200 feet (61 m) maximum. Passing spaces shall be either a 60-inch (1525 mm) minimum by 60-inch (1525 mm) minimum space, or an intersection of two walking surfaces that provide a T-shaped turning space complying with Section 304.3.2, provided the base and arms of the T-shaped space extend 48 inches (1220 mm) minimum beyond the intersection.

403.6 Handrails. Where handrails are required at the side of a corridor they shall comply with Sections 505.4 through 505.9.

404 Doors and Doorways

404.1 General. Doors and doorways that are part of an accessible route shall comply with Section 404.

404.2 Manual Doors. Manual doors and doorways, and manual gates, including ticket gates, shall comply with Section 404.2.

> **EXCEPTION:** Doors, doorways, and gates designed to be operated only by security personnel shall not be required to comply with Sections 404.2.6, 404.2.7, and 404.2.8.

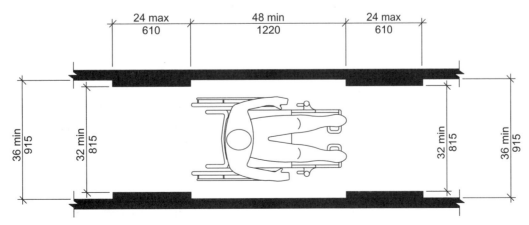

FIG. 403.5
CLEAR WIDTH OF AN ACCESSIBLE ROUTE

404.2.1 Double-Leaf Doors and Gates. At least one of the active leaves of doorways with two leaves shall comply with Sections 404.2.2 and 404.2.3.

404.2.2 Clear Width. Doorways shall have a clear opening width of 32 inches (815 mm) minimum. Clear opening width of doorways with swinging doors shall be measured between the face of door and stop, with the door open 90 degrees. Openings more than 24 inches (610 mm) in depth at doors and doorways without doors shall provide a clear opening width of 36 inches (915 mm) minimum. There shall be no projections into the clear opening width lower

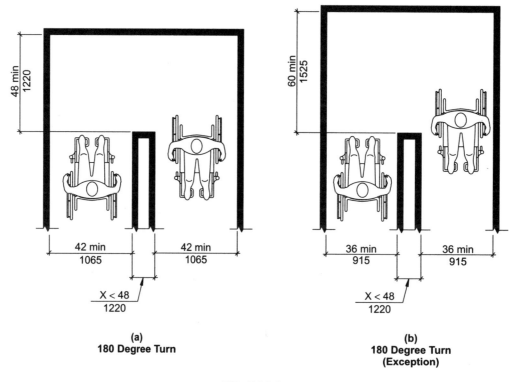

(a)
180 Degree Turn

(b)
**180 Degree Turn
(Exception)**

**FIG. 403.5.1
CLEAR WIDTH AT 180° TURN**

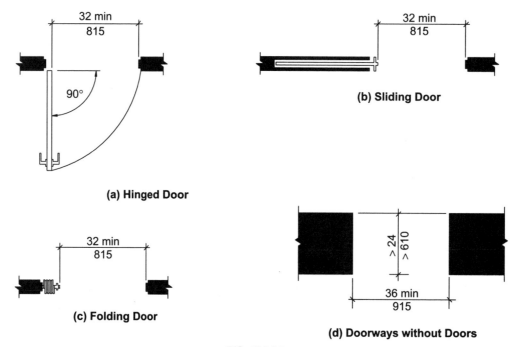

(a) Hinged Door

(b) Sliding Door

(c) Folding Door

(d) Doorways without Doors

**FIG. 404.2.2
CLEAR WIDTH OF DOORWAYS**

than 34 inches (865 mm) above the floor. Projections into the clear opening width between 34 inches (865 mm) and 80 inches (2030 mm) above the floor shall not exceed 4 inches (100 mm).

EXCEPTIONS:

1. Door closers and door stops shall be permitted to be 78 inches (1980 mm) minimum above the floor.

2. In alterations, a projection of $^5/_8$ inch (16 mm) maximum into the required clear opening width shall be permitted for the latch side stop.

404.2.3 Maneuvering Clearances. Minimum maneuvering clearances at doors shall comply with Section 404.2.3 and shall include the full clear opening width of the doorway. Required door maneuvering clearances shall not include knee and toe clearance.

404.2.3.1 Floor Surface. Floor surface within the maneuvering clearances shall have a slope not steeper than 1:48 and shall comply with Section 302.

404.2.3.2 Swinging Doors. Swinging doors shall have maneuvering clearances complying with Table 404.2.3.2.

404.2.3.3 Sliding and Folding Doors. Sliding doors and folding doors shall have maneuvering clearances complying with Table 404.2.3.3.

TABLE 404.2.3.2—MANEUVERING CLEARANCES AT MANUAL SWINGING DOORS

TYPE OF USE		MANEUVERING CLEARANCES AT MANUAL SWINGING DOORS	
Approach Direction	Door Side	Perpendicular to Doorway	Parallel to Doorway (beyond latch unless noted)
From front	Pull	60 inches (1525 mm)	18 inches (455 mm)
From front	Push	48 inches (1220 mm)	0 inches (0 mm)[3]
From hinge side	Pull	60 inches (1525 mm)	36 inches (915 mm)
From hinge side	Pull	54 inches (1370 mm)	42 inches (1065 mm)
From hinge side	Push	42 inches (1065 mm)[1]	22 inches (560 mm)[3 & 4]
From latch side	Pull	48 inches (1220 mm)[1]	24 inches (610 mm)
From latch side	Push	42 inches (1065 mm)[2]	24 inches (610 mm)

[1]Add 6 inches (150 mm) if closer and latch provided.
[2]Add 6 inches (150 mm) if closer provided.
[3]Add 12 inches (305 mm) beyond latch if closer and latch provided.
[4]Beyond hinge side.

TABLE 404.2.3.3—MANEUVERING CLEARANCES AT SLIDING AND FOLDING DOORS

	MINIMUM MANEUVERING CLEARANCES	
Approach Direction	Perpendicular to Doorway	Parallel to Doorway (beyond stop or latch side unless noted)
From front	48 inches (1220 mm)	0 inches (0 mm)
From nonlatch side	42 inches (1065 mm)	22 inches (560 mm)[1]
From latch side	42 inches (1065 mm)	24 inches (610 mm)

[1]Beyond pocket or hinge side.

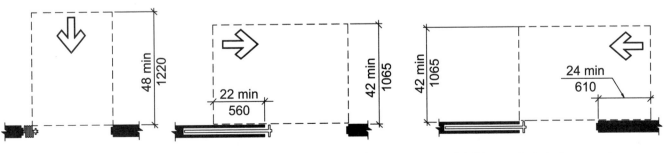

(a) Front Approach   **(b) Pocket or Hinge Approach**   **(c) Stop or Latch Approach**

FIG. 404.2.3.3
MANEUVERING CLEARANCE AT SLIDING AND FOLDING DOORS

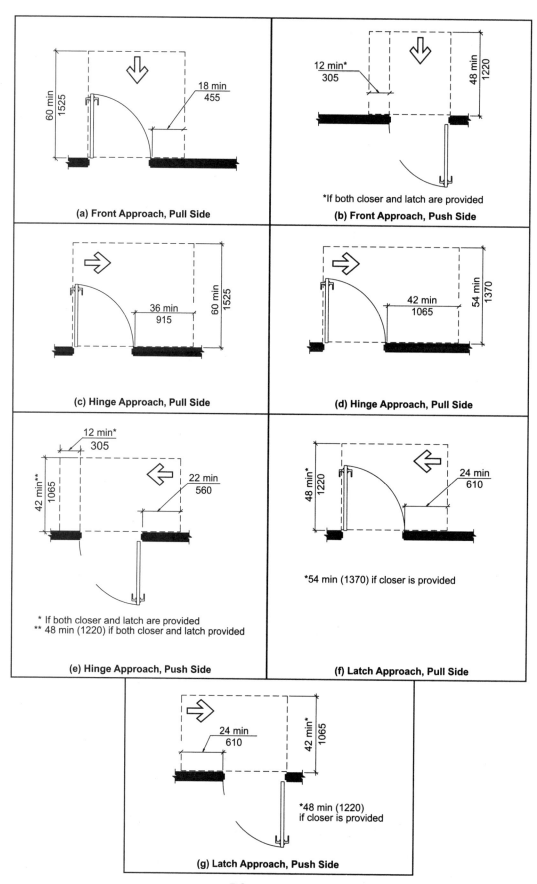

FIG. 404.2.3.2
MANEUVERING CLEARANCE AT MANUAL SWINGING DOORS

404.2.3.4 Doorways without Doors. Doorways without doors that are less than 36 inches (915 mm) in width shall have maneuvering clearances complying with Table 404.2.3.4

404.2.3.5 Recessed Doors. Where any obstruction within 18 inches (455 mm) of the latch side of a doorway projects more than 8 inches (205 mm) beyond the face of the door, measured perpendicular to the face of the door, maneuvering clearances for a forward approach shall be provided.

404.2.4 Thresholds. If provided, thresholds at doorways shall be $^1/_2$ inch (13 mm) maximum in height. Raised thresholds and changes in level at doorways shall comply with Sections 302 and 303.

EXCEPTION: An existing or altered threshold shall be permitted to be $^3/_4$ inch (19 mm) maximum in height provided that the threshold has a beveled edge on each side with a maximum slope of 1:2 for the height exceeding $^1/_4$ inch (6.4 mm).

404.2.5 Two Doors in Series. Distance between two hinged or pivoted doors in series shall be 48 inches (1220 mm) minimum plus the width of any door swinging into the space. The space between the doors shall provide a turning space complying with Section 304.

404.2.6 Door Hardware. Handles, pulls, latches, locks, and other operable parts on accessible doors shall have a shape that is easy to grasp with one hand and does not require tight grasping, pinching, or twisting of the wrist to operate. Operable parts of such hardware shall be 34 inches (865 mm) minimum and 48 inches (1220 mm) maximum above the floor. Where sliding doors are in the fully open position, operating hardware shall be exposed and usable from both sides.

EXCEPTION: Locks used only for security purposes and not used for normal operation shall not be required to comply with Section 404.2.6.

404.2.7 Closing Speed.

404.2.7.1 Door Closers. Door closers shall be adjusted so that from an open position of 90 degrees, the time required to move the door to an open position of 12 degrees shall be 5 seconds minimum.

404.2.7.2 Spring Hinges. Door spring hinges shall be adjusted so that from an open position of 70 degrees, the door shall move to the closed position in 1.5 seconds minimum.

TABLE 404.2.3.4—MANEUVERING CLEARANCES FOR DOORWAYS WITHOUT DOORS

Approach Direction	MINIMUM MANEUVERING CLEARANCES Perpendicular to Doorway
From front	48 inches (1220 mm)
From side	42 inches (1065 mm)

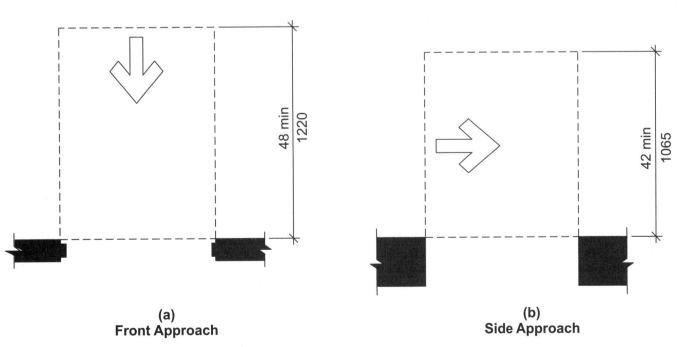

(a)
Front Approach

48 min / 1220

(b)
Side Approach

42 min / 1065

FIG. 404.2.3.4
MANEUVERING CLEARANCE AT DOORWAYS WITHOUT DOORS

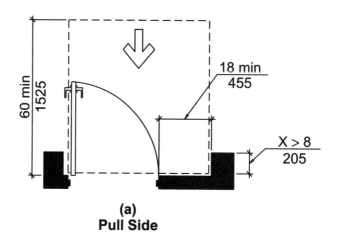

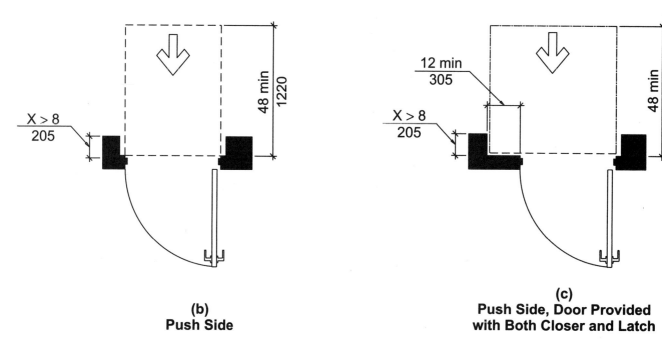

FIG. 404.2.3.5
MANEUVERING CLEARANCE AT RECESSED DOORS

404.2.8 Door-Opening Force. Fire doors shall have the minimum opening force allowable by the appropriate administrative authority. The force for pushing or pulling open doors other than fire doors shall be as follows:

1. Interior hinged door: 5.0 pounds (22.2 N) maximum

2. Sliding or folding door: 5.0 pounds (22.2 N) maximum

These forces do not apply to the force required to retract latch bolts or disengage other devices that hold the door in a closed position.

404.2.9 Door Surface. Door surfaces within 10 inches (255 mm) of the floor, measured vertically, shall be a smooth surface on the push side extending the full width of the door. Parts creating horizontal or vertical joints in such surface shall be within $1/16$ inch (1.6 mm) of the same plane as the other. Cavities created by added kick plates shall be capped.

EXCEPTIONS:

1. Sliding doors shall not be required to comply with Section 404.2.9.

2. Tempered glass doors without stiles and having a bottom rail or shoe with the top leading edge tapered at no less than 60 degrees from the horizontal shall not be required to comply with the 10-inch (255 mm) bottom rail height requirement.

3. Doors that do not extend to within 10 inches (255 mm) of the floor shall not be required to comply with Section 404.2.9.

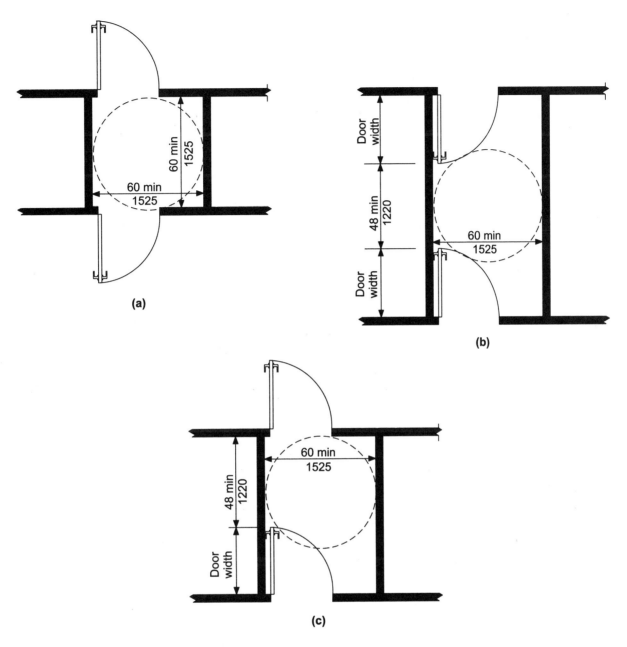

Fig. 404.2.5
TWO DOORS IN A SERIES

404.2.10 Vision Lites. Doors and sidelites adjacent to doors containing one or more glazing panels that permit viewing through the panels shall have the bottom of at least one panel on either the door or an adjacent sidelite 43 inches (1090 mm) maximum above the floor.

> **EXCEPTION:** Vision lites with the lowest part more than 66 inches (1675 mm) above the floor shall not be required to comply with Section 404.2.10.

404.3 Automatic Doors. Automatic doors and automatic gates shall comply with Section 404.3. Full powered automatic doors shall comply with ANSI/BHMA A156.10 listed in Section 105.2.4. Power-assist and low-energy doors shall comply with ANSI/BHMA A156.19 listed in Section 105.2.3.

> **EXCEPTION:** Doors, doorways, and gates designed to be operated only by security personnel shall not be required to comply with Sections 404.3.2, 404.3.4, and 404.3.5.

404.3.1 Clear Width. Doorways shall have a clear opening width of 32 inches (815 mm) in power-on and power-off mode. The minimum clear opening width for automatic door systems shall be based on the clear opening width provided with all leafs in the open position.

404.3.2 Maneuvering Clearances. Maneuvering clearances at power-assisted doors shall comply with Section 404.2.3.

404.3.3 Thresholds. Thresholds and changes in level at doorways shall comply with Section 404.2.4.

404.3.4 Two Doors in Series. Doors in series shall comply with Section 404.2.5.

404.3.5 Control Switches. Manually operated control switches shall comply with Section 309. The clear floor space adjacent to the control switch shall be located beyond the arc of the door swing.

405 Ramps

405.1 General. Ramps along accessible routes shall comply with Section 405.

EXCEPTION: In assembly areas, aisle ramps adjacent to seating and not serving elements required to be on an accessible route shall not be required to comply with Section 405.

405.2 Slope. Ramp runs shall have a running slope greater than 1:20 and not steeper than 1:12.

EXCEPTION: In existing buildings or facilities, ramps shall be permitted to have slopes steeper than 1:12 complying with Table 405.2 where such slopes are necessary due to space limitations.

405.3 Cross Slope. Cross slope of ramp runs shall not be steeper than 1:48.

405.4 Floor Surfaces. Floor surfaces of ramp runs shall comply with Section 302.

405.5 Clear Width. The clear width of a ramp run shall be 36 inches (915 mm) minimum. Handrails and handrail supports that are provided on the ramp run shall not project into the required clear width of the ramp run.

405.6 Rise. The rise for any ramp run shall be 30 inches (760 mm) maximum.

405.7 Landings. Ramps shall have landings at the bottom and top of each ramp run. Landings shall comply with Section 405.7.

405.7.1 Slope. Landings shall have a slope not steeper than 1:48 and shall comply with Section 302.

405.7.2 Width. Clear width of landings shall be at least as wide as the widest ramp run leading to the landing.

405.7.3 Length. Landings shall have a clear length of 60 inches (1525 mm) minimum.

405.7.4 Change in Direction. Ramps that change direction at ramp landings shall be sized to provide a turning space complying with Section 304.3.

405.7.5 Doorways. Where doorways are adjacent to a ramp landing, maneuvering clearances required by Sections 404.2.3 and 404.3.2 shall be permitted to overlap the landing area. Where a door that is subject to locking is located adjacent to a ramp landing, the landing shall be sized to provide a turning space complying with Section 304.3.

405.8 Handrails. Ramp runs with a rise greater than 6 inches (150 mm) shall have handrails complying with Section 505.

TABLE 405.2—ALLOWABLE RAMP DIMENSIONS FOR CONSTRUCTION IN EXISTING SITES, BUILDINGS AND FACILITIES

Slope[1]	Maximum Rise
Steeper than 1:10 but not steeper than 1:8	3 inches (75 mm)
Steeper than 1:12 but not steeper than 1:10	6 inches (150 mm)

[1]A slope steeper than 1:8 shall not be permitted.

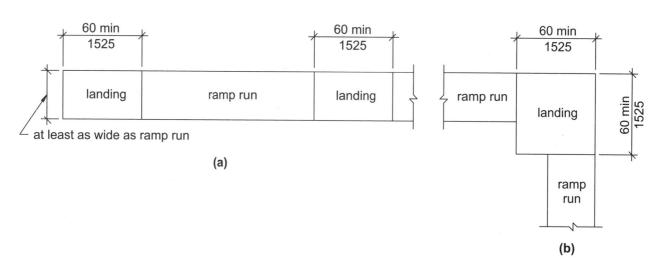

(a)

(b)

FIG. 405.7
RAMP LANDINGS

405.9 Edge Protection. Edge protection complying with Section 405.9.1 or 405.9.2 shall be provided on each side of ramp runs and at each side of ramp landings.

EXCEPTIONS:

1. Edge protection shall not be required on ramps not required to have handrails and that have flared sides complying with Section 406.3.

2. Edge protection shall not be required on the sides of ramp landings serving an adjoining ramp run or stairway.

3. Edge protection shall not be required on the sides of ramp landings having a vertical drop-off of $^1/_2$ inch (13 mm) maximum within 10 inches (255 mm) horizontally of the minimum landing area specified in Section 405.7.

4. Edge protection shall not be required on the sides of ramped aisles where the ramps provide access to the adjacent seats and aisle access ways.

405.9.1 Extended Floor Surface. The floor surface of the ramp run or ramp landing shall extend 12 inches (305 mm) minimum beyond the inside face of a railing complying with Section 505.

405.9.2 Curb or Barrier. A curb complying with Section 405.9.2.1 or a barrier complying with Section 405.9.2.2 shall be provided.

405.9.2.1 Curb. A curb shall be a minimum of 4 inches (100 mm) in height.

405.9.2.2 Barrier. Barriers shall be constructed so that the barrier prevents the passage of a 4-inch (100 mm) diameter sphere where any portion of the sphere is within 4 inches (100 mm) of the floor.

405.10 Wet Conditions. Landings subject to wet conditions shall be designed to prevent the accumulation of water.

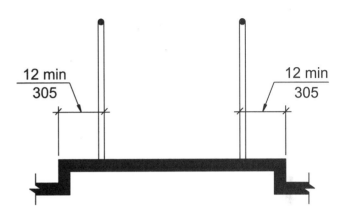

Extended Floor Surface

FIG. 405.9.1
EXTENDED FLOOR SURFACE

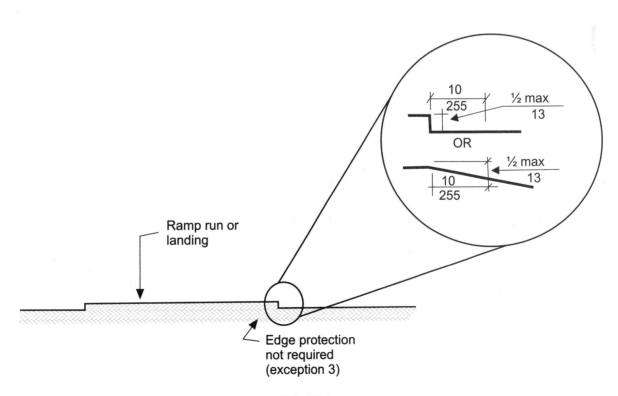

FIG. 405.9
EDGE PROTECTION—LIMITED DROP OFF

406 Curb Ramps

406.1 General. Curb ramps on accessible routes shall comply with Sections 406, 405.2, 405.3, and 405.10.

406.2 Counter Slope. Counter slopes of adjoining gutters and road surfaces immediately adjacent to the curb ramp shall not be steeper than 1:20. The adjacent surfaces at transitions at curb ramps to walks, gutters and streets shall be at the same level.

406.3 Sides of Curb Ramps. Where provided, curb ramp flares shall comply with Section 406.3.

406.3.1 Slope. Flares shall not be steeper than 1:10.

406.3.2 Marking. If curbs adjacent to the ramp flares are painted, the painted surface shall extend along the flared portion of the curb.

406.4 Width. Curb ramps shall be 36 inches (915 mm) minimum in width, exclusive of flared sides.

406.5 Floor Surface. Floor surfaces of curb ramps shall comply with Section 302.

406.6 Location. Curb ramps and the flared sides of curb ramps shall be located so they do not project into vehicular traffic lanes, parking spaces, or parking access aisles. Curb ramps at marked crossings shall be wholly contained within the markings, excluding any flared sides.

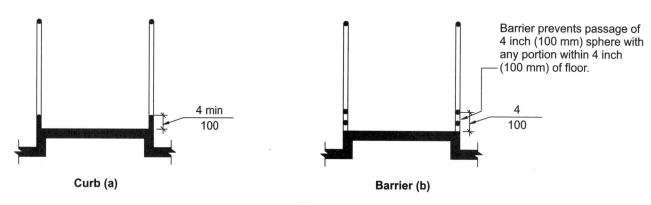

FIG. 405.9.2
RAMP EDGE PROTECTION

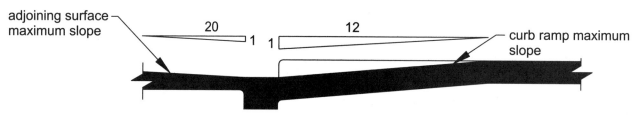

FIG. 406.2
COUNTER SLOPE OF SURFACES ADJACENT TO CURB RAMPS

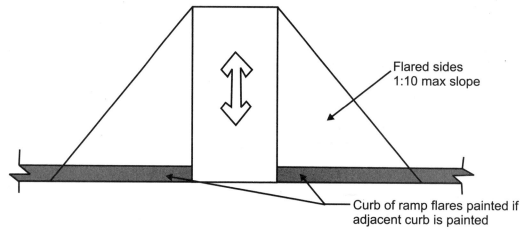

FIG. 406.3
SIDES OF CURB RAMPS

406.7 Landings. Landings shall be provided at the tops of curb ramps. The clear length of the landing shall be 36 inches (915 mm) minimum. The clear width of the landing shall be at least as wide as the curb ramp, excluding flared sides, leading to the landing.

EXCEPTION: In alterations, where there is no landing at the top of curb ramps, curb ramp flares shall be provided and shall not be steeper than 1:12.

406.8 Obstructions. Curb ramps shall be located or protected to prevent their obstruction by parked vehicles.

406.9 Handrails. Handrails shall not be required on curb ramps.

406.10 Diagonal Curb Ramps. Diagonal or corner-type curb ramps with returned curbs or other well-defined edges shall have the edges parallel to the direction of pedestrian flow. The bottoms of diagonal curb ramps shall have 48 inches (1220 mm) minimum clear space outside active traffic lanes of the roadway. Diagonal curb ramps provided at marked crossings shall provide the 48 inches (1220 mm) minimum clear space within the markings. Diagonal curb ramps with flared sides shall have a segment of curb 24 inches (610 mm) minimum in length on each side of the curb ramp and within the marked crossing.

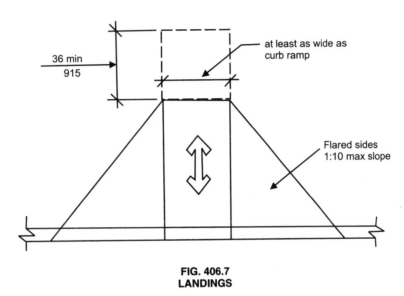

FIG. 406.7
LANDINGS

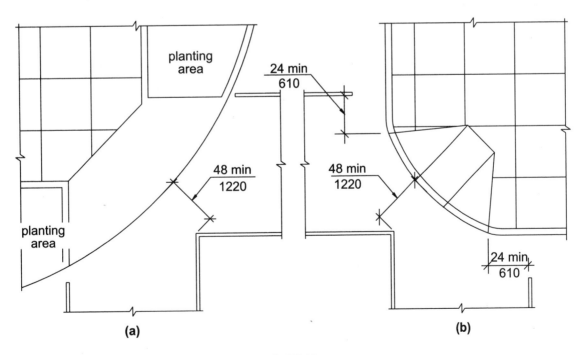

(a)

(b)

FIG. 406.10
DIAGONAL CURB RAMPS

406.11 Islands. Raised islands in crossings shall be a cut-through level with the street or have curb ramps at both sides. Each curb ramp shall have a level area 48 inches (1220 mm) minimum in length and 36 inches (915 mm) minimum in width at the top of the curb ramp in the part of the island intersected by the crossings. Each 48-inch (1220 mm) by 36-inch (915 mm) area shall be oriented so the 48-inch (1220 mm) length is in the direction of the running slope of the curb ramp it serves. The 48-inch (1220 mm) by 36-inch (915 mm) areas and the accessible route shall be permitted to overlap.

406.12 Detectable Warnings at Raised Marked Crossings. Marked crossings that are raised to the same level as the adjoining sidewalk shall be preceded by a detectable warning 24 inches (610 mm) in depth complying with Section 705. The detectable warning shall extend the full width of the marked crossing.

406.13 Detectable Warnings at Curb Ramps. Where detectable warnings are provided on curb ramps, they shall comply with Sections 406.13 and 705.

406.13.1 Area Covered. Detectable warnings shall be 24 inches (610 mm) minimum in depth in the direction of travel. The detectable warning shall extend the full width of the curb ramp or flush surface.

406.13.2 Location. The detectable warning shall be located so the edge nearest the curb line is 6 inches (150 mm) minimum and 8 inches (205 mm) maximum from the curb line.

406.14 Detectable Warnings at Islands or Cut-through Medians. Where detectable warnings are provided on curb ramps or at raised marked crossings leading to islands or cut-through medians, the island or cut-through median shall be provided with detectable warnings complying with Section 705, that are 24 inches (610 mm) in depth, and extend the full width of the pedestrian route or cut-through. Where such island or cut-through median is less than 48 inches (1220 mm) in depth, the entire width and depth of the pedestrian route or cut-through shall have detectable warnings.

407 Elevators

407.1 General. Elevators shall comply with Section 407 and ASME A17.1/CSA B44 listed in Section 105.2.5. Elevators shall be passenger elevators as classified by ASME A17.1/CSA B44. Elevator operation shall be automatic.

407.2 Elevator Landing Requirements. Elevator landings shall comply with Section 407.2.

407.2.1 Call Controls. Where elevator call buttons or keypads are provided, they shall comply with Sections 407.2.1 and 309.4. Call buttons shall be raised or flush. Objects beneath hall call buttons shall protrude 1 inch (25 mm) maximum.

EXCEPTIONS:

1. Existing elevators shall be permitted to have recessed call buttons.

2. The restriction on objects beneath call buttons shall not apply to existing call buttons.

407.2.1.1 Height. Call buttons and keypads shall be located within one of the reach ranges speci-

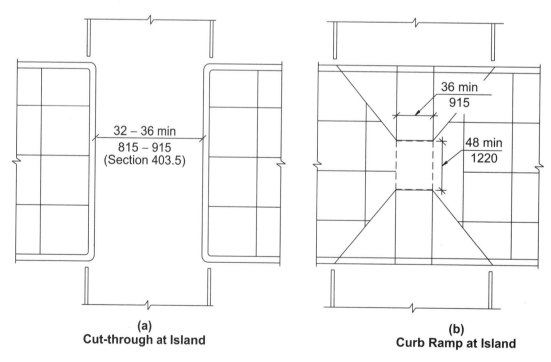

(a)
Cut-through at Island

(b)
Curb Ramp at Island

FIG. 406.11
ISLANDS

fied in Section 308, measured to the centerline of the highest operable part.

EXCEPTION: Existing call buttons and existing keypads shall be permitted to be located 54 inches (1370 mm) maximum above the floor, measured to the centerline of the highest operable part.

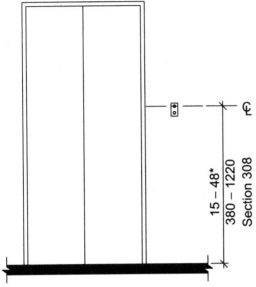

*54 max (1370) for existing

**FIG. 407.2.1.1
HEIGHT OF ELEVATOR CALL BUTTONS**

407.2.1.2 Size. Call buttons shall be $^3/_4$ inch (19 mm) minimum in the smallest dimension.

EXCEPTION: Existing elevator call buttons shall not be required to comply with Section 407.2.1.2.

407.2.1.3 Clear Floor Space. A clear floor space complying with Section 305 shall be provided at call controls.

407.2.1.4 Location. The call button that designates the up direction shall be located above the call button that designates the down direction.

EXCEPTION: Destination-oriented elevators shall not be required to comply with Section 407.2.1.4.

407.2.1.5 Signals. Call buttons shall have visible signals to indicate when each call is registered and when each call is answered. Call buttons shall provide an audible signal or mechanical motion of the button to indicate when each call is registered.

EXCEPTIONS:

1. Destination-oriented elevators shall not be required to comply with Section 407.2.1.5, provided a visible signal and audible tones and verbal announcements complying with Section 407.2.1.7 are provided.

2. Existing elevators shall not be required to comply with Section 407.2.1.5.

407.2.1.6 Keypads. Where keypads are provided, keypads shall be in a standard telephone keypad arrangement and shall comply with Section 407.4.7.2.

407.2.1.7 Destination-oriented Elevator Signals. Destination-oriented elevators shall be provided with a visible signal and audible tones and verbal announcements to indicate which car is responding to a call. The audible tone and verbal announcement shall be activated by pressing a function button. The function button shall be identified by the International Symbol for Accessibility and a raised indication. The International Symbol for Accessibility, complying with Section 703.6.3.1, shall be $^5/_8$ inch (16 mm) in height and be a visual character complying with Section 703.2. The indication shall be three raised dots, spaced $^1/_4$ inch (6.4 mm) at base diameter, in the form of an equilateral triangle. The function button shall be located immediately below the keypad arrangement or floor buttons.

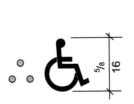

Visual and raised information

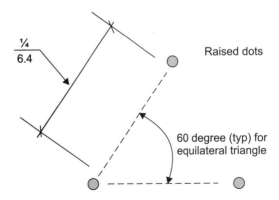

Raised information

**FIG. 407.2.1.7
DESTINATION-ORIENTED ELEVATOR INDICATION**

407.2.2 Hall Signals. Hall signals, including in-car signals, shall comply with Section 407.2.2.

407.2.2.1 Visible and Audible Signals. A visible and audible signal shall be provided at each hoistway entrance to indicate which car is answering a call and the car's direction of travel. Where in-car signals are provided they shall be visible from the floor area adjacent to the hall call buttons.

EXCEPTIONS:

1. Destination-oriented elevators shall not be required to comply with Section 407.2.2.1, provided a visible signal and audible tones and verbal announcements complying with Section 407.2.1.7 are provided.

2. In existing elevators, a signal indicating the direction of car travel shall not be required.

407.2.2.2 Visible Signals. Visible signal fixtures shall be centered at 72 inches (1830 mm) minimum above the floor. The visible signal elements shall be 2$^1/_2$ inches (64 mm) minimum between the uppermost and lowest edges of the illuminated shape measured vertically. Signals shall be visible from the floor area adjacent to the hall call button.

EXCEPTIONS:

1. Destination-oriented elevators shall be permitted to have signals visible from the floor area adjacent to the hoistway entrance.

2. Existing elevators shall not be required to comply with Section 407.2.2.2.

407.2.2.3 Audible Signals. Audible signals shall sound once for the up direction and twice for the down direction, or shall have verbal annunciators that indicate the direction of elevator car travel. Audible signals shall have a frequency of 1500 Hz maximum. Verbal annunciators shall have a frequency of 300 Hz minimum and 3,000 Hz maximum. The audible signal or verbal annunciator shall be 10 dBA minimum above ambient, but shall not exceed 80 dBA, measured at the hall call button.

EXCEPTIONS:

1. Destination-oriented elevators shall not be required to comply with Section 407.2.2.3, provided the audible tone and verbal announcement is the same as those given at the call button or call button keypad.

2. The requirement for the frequency and range of audible signals shall not apply in existing elevators.

407.2.2.4 Differentiation. Each destination-oriented elevator in a bank of elevators shall have audible and visible means for differentiation.

407.2.3 Hoistway Signs. Signs at elevator hoistways shall comply with Section 407.2.3.

407.2.3.1 Floor Designation. Floor designations shall be provided in raised characters and braille complying with Sections 703.3 and 703.4. Raised characters shall be 2 inches (51 mm) minimum in height. Floor designations shall be located on both jambs of elevator hoistway entrances. A raised star shall be provided on both jambs at the main entry level.

407.2.3.2 Car Identification. Destination-oriented elevators shall provide car identification in raised characters and braille complying with Sec-

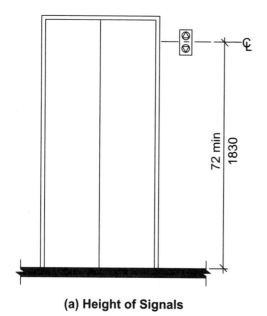

72 min
1830

(a) Height of Signals

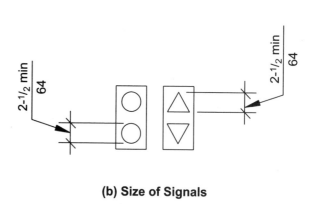

2-1/2 min
64

2-1/2 min
64

(b) Size of Signals

FIG. 407.2.2.2
ELEVATOR VISIBLE SIGNALS

tions 703.3 and 703.4. Raised characters shall be 2 inches (51 mm) minimum in height. Car identifications shall be located on both jambs of the hoistway immediately below the floor designation.

407.2.4 Destination Signs. Where signs indicate that elevators do not serve all landings, signs in raised characters and braille complying with Sections 703.3 and 703.4 shall be provided above the hall call button or keypad.

EXCEPTION: Destination oriented elevator systems shall not be required to comply with Section 407.2.4.

407.3 Elevator Door Requirements. Hoistway and elevator car doors shall comply with Section 407.3.

407.3.1 Type. Elevator doors shall be horizontal sliding type. Car gates shall be prohibited.

407.3.2 Operation. Elevator hoistway and car doors shall open and close automatically.

EXCEPTION: Existing manually operated hoistway swing doors shall be permitted, provided the following criteria are met:

a) The hoistway doors comply with Sections 404.2.2 and 404.2.8;

b) The car door closing is not initiated until the hoistway door is closed.

407.3.3 Reopening Device. Elevator doors shall be provided with a reopening device complying with Section 407.3.3 that shall stop and reopen a car door and hoistway door automatically if the door becomes obstructed by an object or person.

EXCEPTION: In existing elevators, manually operated doors shall not be required to comply with Section 407.3.3.

407.3.3.1 Height. The reopening device shall be activated by sensing an obstruction passing through the opening at 5 inches (125 mm) nominal and 29 inches (735 mm) nominal above the floor.

407.3.3.2 Contact. The reopening device shall not require physical contact to be activated, although contact shall be permitted before the door reverses.

407.3.3.3 Duration. The reopening device shall remain effective for 20 seconds minimum.

407.3.4 Door and Signal Timing. The minimum acceptable time from notification that a car is answering a call until the doors of that car start to close shall be calculated from the following equation:

$T = D/(1.5 \text{ ft/s})$ or $T = D/(455 \text{ mm/s}) = 5$ seconds minimum, where T equals the total time in seconds and D equals the distance (in feet or millimeters) from the point in the lobby or corridor 60 inches (1525 mm) directly in front of the farthest call button controlling that car to the centerline of its hoistway door.

EXCEPTIONS:

1. For cars with in-car lanterns, T shall be permitted to begin when the signal is visible from the point 60 inches (1525 mm) directly in front of the farthest hall call button and the audible signal is sounded.

2. Destination-oriented elevators shall not be required to comply with Section 407.3.4.

407.3.5 Door Delay. Elevator doors shall remain fully open in response to a car call for 3 seconds minimum.

407.3.6 Width. Elevator door clear opening width shall comply with Table 407.4.1.

EXCEPTION: In existing elevators, a power-operated car door complying with Section 404.2.2 shall be permitted.

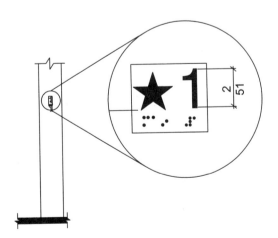

**FIG. 407.2.3.1
FLOOR DESIGNATION**

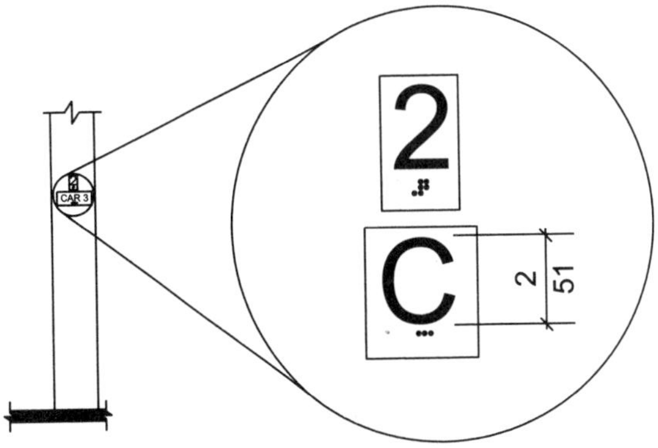

**FIG. 407.2.3.2
DESTINATION-ORIENTED ELEVATOR CAR IDENTIFICATION**

407.4 Elevator Car Requirements. Elevator cars shall comply with Section 407.4.

407.4.1 Inside Dimensions. Inside dimensions of elevator cars shall comply with Table 407.4.1.

EXCEPTION: Existing elevator car configurations that provide a clear floor area of 16 square feet (1.5 m²) minimum, and provide a clear inside dimension of 36 inches (915 mm) minimum in width and 54 inches (1370 mm) minimum in depth, shall be permitted.

407.4.2 Floor Surfaces. Floor surfaces in elevator cars shall comply with Section 302.

407.4.3 Platform to Hoistway Clearance. The clearance between the car platform sill and the edge of any hoistway landing shall comply with ASME A17.1/CSA B44 listed in Section 105.2.5.

407.4.4 Leveling. Each car shall automatically stop and maintain position at floor landings within a tolerance of ¹/₂ inch (13 mm) under rated loading to zero loading conditions.

407.4.5 Illumination. The level of illumination at the car controls, platform, car threshold and car landing sill shall comply with ASME A17.1/CSA B44 listed in Section 105.2.5.

407.4.6 Elevator Car Controls. Where provided, elevator car controls shall comply with Sections 407.4.6 and 309.

EXCEPTION: In existing elevators, where a new car operating panel complying with Section 407.4.6 is provided, existing car operating panels shall not be required to comply with Section 407.4.6.

407.4.6.1 Location. Controls shall be located within one of the reach ranges specified in Section 308.

EXCEPTIONS:

1. Where the elevator panel complies with Section 407.4.8.

2. In existing elevators, where a parallel approach is provided to the controls, car control buttons with floor designations shall be permitted to be located 54 inches (1370 mm) maximum above the floor. Where the panel is changed, it shall comply with Section 407.4.6.1.

407.4.6.2 Buttons. Car control buttons with floor designations shall be raised or flush, and shall comply with Section 407.4.6.2.

EXCEPTION: In existing elevators, buttons shall be permitted to be recessed.

407.4.6.2.1 Size. Buttons shall be ³/₄ inch (19 mm) minimum in their smallest dimension.

407.4.6.2.2 Arrangement. Buttons shall be arranged with numbers in ascending order. Floors shall be designated . . . -4, -3, -2, -1, 0, 1, 2, 3, 4, etcetera, with floors below the main entry floor designated with minus numbers. Numbers shall be permitted to be omitted, provided the remaining numbers are in sequence. Where a telephone keypad arrangement is used, the number key ("#") shall be utilized to enter the minus symbol ("-"). When two or more columns of buttons are provided they shall read from left to right.

407.4.6.3 Keypads. Where provided, car control keypads shall be in a standard telephone keypad arrangement and shall comply with Section 407.4.7.2.

407.4.6.4 Emergency Controls. Emergency controls shall comply with Section 407.4.6.4.

407.4.6.4.1 Height. Emergency control buttons shall have their centerlines 35 inches (890 mm) minimum above the floor.

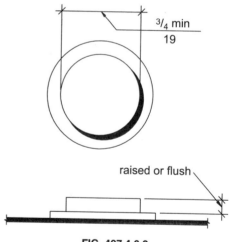

³/₄ min
19

raised or flush

**FIG. 407.4.6.2
ELEVATOR CAR CONTROL BUTTONS**

TABLE 407.4.1—MINIMUM DIMENSIONS OF ELEVATOR CARS

Door Location	Door Clear Opening Width	Inside Car, Side to Side	Inside Car, Back Wall to Front Return	Inside Car, Back Wall to Inside Face
Centered	42 inches (1065 mm)	80 inches (2030 mm)	51 inches (1295 mm)	54 inches (1370 mm)
Side (Off Center)	36 inches (915 mm)[1]	68 inches (1725 mm)	51 inches (1295 mm)	54 inches (1370 mm)
Any	36 inches (915 mm)[1]	54 inches (1370 mm)	80 inches (2030 mm)	80 inches (2030 mm)
Any	36 inches (915 mm)[1]	60 inches (1525 mm)[2]	60 inches (1525 mm)[2]	60 inches (1525 mm)[2]

[1]A tolerance of minus ⁵/₈ inch (16 mm) is permitted.

[2]Other car configurations that provide a 36-inch (915 mm) door clear opening width and a turning space complying with Section 304 with the door closed are permitted.

407.4.6.4.2 Location. Emergency controls, including the emergency alarm, shall be grouped at the bottom of the panel.

407.4.7 Designations and Indicators of Car Controls. Designations and indicators of car controls shall comply with Section 407.4.7.

EXCEPTIONS:

1. In existing elevators, where a new car operating panel complying with Section 407.4.7 is provided, existing car operating panels shall not be required to comply with Section 407.4.7.

2. Where existing building floor designations differ from the arrangement required by Sec-

tion 407.4.6.2.2, or are alphanumeric, a new operating panel shall be permitted to use such existing building floor designations.

407.4.7.1 Buttons. Car control buttons shall comply with Section 407.4.7.1.

407.4.7.1.1 Type. Control buttons shall be identified by raised characters and braille complying with Sections 703.3 and 703.4.

407.4.7.1.2 Location. Raised character and braille designations shall be placed immediately to the left of the control button to which the designations apply. Where a negative number is used to indicate a negative floor, the braille

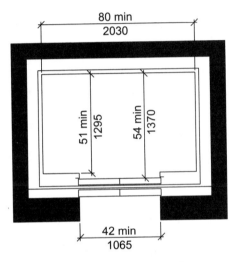

(a) Centered Door Location

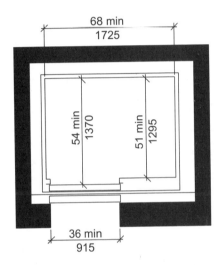

(b) Side (Off-Centered Door) Location

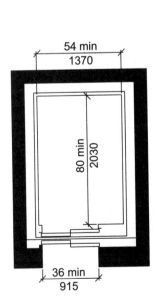

(c) Any Door Location

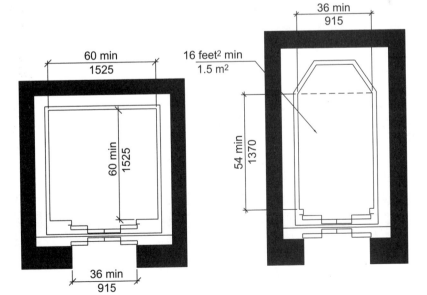

(d) Any Door Location　　　　　　**(e) Existing Car Configurations**

FIG. 407.4.1
INSIDE DIMENSIONS OF ELEVATOR CARS

designation shall be a cell with the dots 3 and 6 followed by the ordinal number.

> **EXCEPTION:** Where space on an existing car operating panel precludes raised characters and braille to the left of the control button, markings shall be placed as near to the control button as possible.

407.4.7.1.3 Symbols. The control button for the emergency stop, alarm, door open, door close, main entry floor, and phone, shall be identified with raised symbols and braille as shown in Table 407.4.7.1.3.

407.4.7.1.4 Visible Indicators. Buttons with floor designations shall be provided with visible indicators to show that a call has been registered. The visible indication shall extinguish when the car arrives at the designated floor.

407.4.7.2 Keypads. Keypad keys shall be identified by visual characters complying with Section 703.2 centered on the corresponding keypad button. The number five key shall have a single raised dot. The dot shall have a base diameter of 0.118 inch (3 mm) minimum and 0.120 inch (3.05 mm) maximum, and a height of 0.025 inch (0.6 mm) minimum and 0.037 inch (0.9 mm) maximum.

407.4.8 Elevator Car Call Sequential Step Scanning. Elevator car call sequential step scanning shall be provided where car control buttons are provided more than 48 inches (1220 mm) above the floor. Floor selection shall be accomplished by applying momentary or constant pressure to the up or down scan button. The up scan button shall sequentially select floors above the current floor. The down scan button shall sequentially select floors below the current floor. When pressure is removed from the up or down scan button for more than 2 seconds, the last floor selected shall be registered as a car call. The up and down scan button shall be located adjacent to or immediately above the emergency control buttons.

407.4.9 Car Position Indicators. Audible and visible car position indicators shall be provided in elevator cars.

407.4.9.1 Visible Indicators. Visible indicators shall comply with Section 407.4.9.1.

407.4.9.1.1 Size. Characters shall be $1/2$ inch (13 mm) minimum in height.

407.4.9.1.2 Location. Indicators shall be located above the car control panel or above the door.

407.4.9.1.3 Floor Arrival. As the car passes a floor and when a car stops at a floor served by the elevator, the corresponding character shall illuminate.

> **EXCEPTION:** Destination-oriented elevators shall not be required to comply with Section 407.4.9.1.3, provided the visible

indicators extinguish when the call has been answered.

407.4.9.1.4 Destination Indicator. In destination-oriented elevators, a display shall be provided in the car with visible indicators to show car destinations.

407.4.9.2 Audible Indicators. Audible indicators shall comply with Section 407.4.9.2.

407.4.9.2.1 Signal Type. The signal shall be an automatic verbal annunciator that announces the floor at which the car is about to stop. The verbal announcement indicating the floor shall be completed prior to the initiation of the door opening.

> **EXCEPTION:** For elevators other than destination-oriented elevators that have a rated speed of 200 feet per minute (1 m/s) maximum, a non-verbal audible signal with a frequency of 1500 Hz maximum that sounds as the car passes or is about to stop at a floor served by the elevator shall be permitted.

407.4.9.2.2 Signal Level. The verbal annunciator shall be 10 dBA minimum above ambient, but shall not exceed 80 dBA, measured at the annunciator.

407.4.9.2.3 Frequency. The verbal annunciator shall have a frequency of 300 Hz minimum and 3,000 Hz maximum.

407.4.10 Emergency Communications. Emergency two-way communication systems between the elevator car and a point outside the hoistway shall comply with Section 407.4.10 and ASME A17.1/CSA B44 listed in Section 105.2.5.

407.4.10.1 Height. The highest operable part of a two-way communication system shall comply with Section 308.

407.4.10.2 Identification. Raised characters and braille complying with Sections 703.3 and 703.4 and raised symbols complying with Section 407.4.7.1.3 shall be provided adjacent to the device.

408 Limited-use/Limited-application Elevators

408.1 General. Limited-use/limited-application elevators shall comply with Section 408 and ASME A17.1/CSA B44 listed in Section 105.2.5. Elevator operation shall be automatic.

408.2 Elevator Landing Requirements. Landings serving limited-use/limited application elevators shall comply with Section 408.2.

408.2.1 Call Controls. Elevator call buttons and keypads shall comply with Section 407.2.1.

408.2.2 Hall Signals. Hall signals shall comply with Section 407.2.2.

408.2.3 Hoistway Signs. Signs at elevator hoistways shall comply with Section 407.2.3.

TABLE 407.4.7.1.3—CONTROL BUTTON IDENTIFICATION

Control Button Type	Raised Symbol	Braille Message	Proportions (Open circles indicate unused dots within each braille cell)
DOOR OPEN 16.0 mm, 4.8 mm, 3.0 mm TYP. BETWEEN ELEMENTS		op"en"	2.0 mm, 2.0 mm
REAR/SIDE DOOR OPEN		op"en"	
DOOR CLOSE		close	
REAR/SIDE DOOR CLOSE		close	
MAIN		ma"in"	
ALARM		al"ar"m	
PHONE		ph"one"	
EMERGENCY STOP (WHEN PROVIDED) X on face of octagon is not required to be tactile		"st"op	

408.3 Elevator Door Requirements. Elevator hoistway doors shall comply with Section 408.3.

408.3.1 Sliding Doors. Sliding hoistway and car doors shall comply with Sections 407.3.1 through 407.3.3, and 408.3.3.

408.3.2 Swinging Doors. Swinging hoistway doors shall open and close automatically and shall comply with Sections 408.3.2, 404, and 407.3.2.

408.3.2.1 Power Operation. Swinging doors shall be power-operated and shall comply with ANSI/BHMA A156.19 listed in Section 105.2.3.

408.3.2.2 Duration. Power-operated swinging doors shall remain open for 20 seconds minimum when activated.

408.3.3 Door Location and Width. Car doors shall comply with Section 408.3.3.

408.3.3.1 Cars with Single Door or Doors on Opposite Ends. Car doors shall be positioned at the narrow end of cars with a single door and on cars with doors on opposite ends. Doors shall provide a clear opening width of 32 inches (815 mm) minimum.

408.3.3.2 Cars with Doors on Adjacent Sides. Car doors shall be permitted to be located on adjacent sides of cars that provide an 18 square foot (1.67 m²) platform. Doors located on the narrow end of cars shall provide a clear opening width of 36 inches (915 mm) minimum. Doors located on the long side shall provide a clear opening width of 42 inches (1065 mm) minimum and be located as far as practicable from the door on the narrow end.

EXCEPTION: Car doors that provide a clear opening width of 36 inches (915 mm) minimum shall be permitted to be located on adjacent sides of cars that provide a clear floor area of 51 inches (1295 mm) in width and 51 inches (1295 mm) in depth.

408.4 Elevator Car Requirements. Elevator cars shall comply with Section 408.4.

408.4.1 Inside Dimensions. Elevator cars shall provide a clear floor width of 42 inches (1065 mm) minimum. The clear floor area shall not be less than 15.75 square feet (1.46 m²).

EXCEPTION: For installations in existing buildings, elevator cars that provide a clear floor area of 15 square feet (1.4 m²) minimum, and provide a clear inside dimension of 36 inches (915 mm) minimum in width and 54 inches (1370 mm) minimum in depth, shall be permitted. This exception shall not apply to cars with doors on adjacent sides.

408.4.2 Floor Surfaces. Floor surfaces in elevator cars shall comply with Section 302.

408.4.3 Platform to Hoistway Clearance. The clearance between the car platform sill and the edge

of any hoistway landing shall comply with ASME A17.1/CSA B44 listed in Section 105.2.5.

408.4.4 Leveling. Elevator car leveling shall comply with Section 407.4.4.

408.4.5 Illumination. Elevator car illumination shall comply with Section 407.4.5.

408.4.6 Elevator Car Controls. Elevator car controls shall comply with Section 407.4.6. Control panels shall be centered on a side wall.

408.4.7 Designations and Indicators of Car Controls. Designations and indicators of car controls shall comply with Section 407.4.7.

408.4.8 Emergency Communications. Car emergency signaling devices complying with Section 407.4.10 shall be provided.

409 Private Residence Elevators

409.1 General. Private residence elevators shall comply with Section 409 and ASME A17.1/CSA B44 listed in Section 105.2.5. Elevator operation shall be automatic.

EXCEPTION: Elevators complying with Section 407 or 408 shall not be required to comply with Section 409.

409.2 Call Controls. Call buttons at elevator landings shall comply with Section 309. Call buttons shall be $^3/_4$ inch (19 mm) minimum in their smallest dimension.

409.3 Doors and Gates. Elevator car and hoistway doors and gates shall comply with Sections 409.3 and 404.

EXCEPTION: The maneuvering clearances required by Section 404.2.3 shall not apply for approaches to the push side of swinging doors.

409.3.1 Power Operation. Elevator car doors and gates shall be power operated and shall comply with ANSI/BHMA A156.19 listed in Section 105.2.3. Elevator cars with a single opening shall have low energy power operated hoistway doors and gates.

EXCEPTION: Hoistway doors or gates shall be permitted to be of the self-closing, manual type, where that door or gate provides access to a narrow end of the car that serves only one landing.

409.3.2 Duration. Power operated doors and gates shall remain open for 20 seconds minimum when activated.

409.3.3 Door or Gate Location and Width. Car gates or doors positioned at a narrow end of the clear floor area required by Section 409.4.1 shall provide a clear opening width of 32 inches (815 mm) minimum. Car gates or doors positioned on adjacent sides shall provide a clear opening width of 42 inches (1065 mm) minimum.

409.4 Elevator Car Requirements. Elevator cars shall comply with Section 409.4.

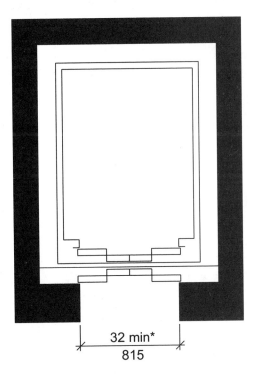

(a) Car with Single Door

*Door opening size from Section 408.3.3

32 min*
815

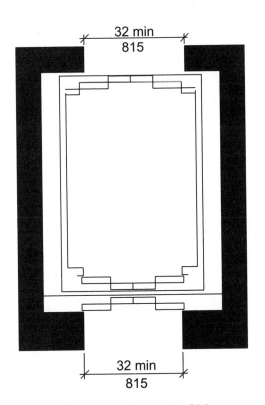

32 min
815

32 min
815

(b) Doors on Opposite Sides

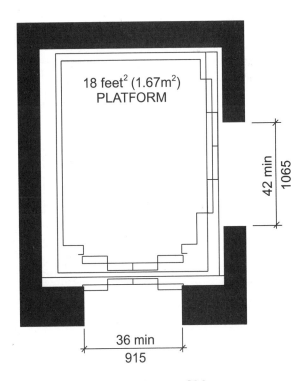

18 feet^2 (1.67m^2)
PLATFORM

42 min
1065

36 min
915

(c) Doors on Adjacent Sides

FIG. 408.3.3
DOOR LOCATION FOR LIMITED USE/LIMITED APPLICATION (LULA) ELEVATORS

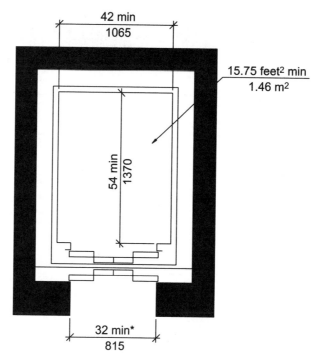

(a) New Construction

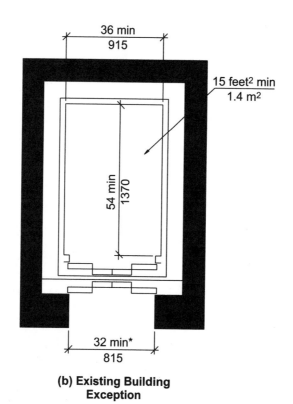

**(b) Existing Building
Exception**

*Door opening size from Section 408.3.3

**FIG. 408.4.1
INSIDE DIMENSIONS OF LIMITED USE/LIMITED APPLICATION (LULA) ELEVATOR CARS**

409.4.1 Inside Dimensions. Elevator cars shall provide a clear floor area 36 inches (915 mm) minimum in width and 48 inches (1220 mm) minimum in depth.

409.4.2 Floor Surfaces. Floor surfaces in elevator cars shall comply with Section 302.

409.4.3 Platform to Hoistway Clearance. The clearance between the car platform sill and the edge of any hoistway landing shall be $1^1/_4$ inches (32 mm) maximum.

409.4.4 Leveling. Each car shall automatically stop at a floor landing within a tolerance of $^1/_2$ inch (13 mm) under rated loading to zero loading conditions.

409.4.5 Illumination. The level of illumination at the car controls, platform, and car threshold and landing sill shall be 5 foot-candles (54 lux) minimum.

409.4.6 Elevator Car Controls. Elevator car controls shall comply with Sections 409.4.6 and 309.4.

409.4.6.1 Buttons. Control buttons shall be $^3/_4$ inch (19 mm) minimum in their smallest dimension. Control buttons shall be raised or flush.

409.4.6.2 Height. Buttons with floor designations shall comply with Section 309.3.

409.4.6.3 Location. Controls shall be on a side-wall, 12 inches (305 mm) minimum from any adjacent wall.

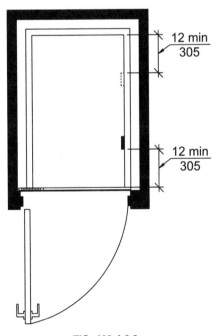

**FIG. 409.4.6.3
LOCATION OF CONTROLS IN
PRIVATE RESIDENCE ELEVATORS**

409.4.7 Emergency Communications. Emergency communications systems shall comply with Section 409.4.7.

409.4.7.1 Type. A telephone and emergency signal device shall be provided in the car.

409.4.7.2 Operable Parts. The telephone and emergency signaling device shall comply with Section 309.3 and 309.4.

409.4.7.3 Compartment. If the device is in a closed compartment, the compartment door hardware shall comply with Section 309.

409.4.7.4 Cord. The telephone cord shall be 29 inches (735 mm) minimum in length.

410 Platform Lifts

410.1 General. Platform lifts shall comply with Section 410 and ASME A18.1 listed in Section 105.2.6. Platform lifts shall not be attendant operated and shall provide unassisted entry and exit from the lift.

410.2 Lift Entry. Lifts with doors or gates shall comply with Section 410.2.1. Lifts with ramps shall comply with Section 410.2.2.

410.2.1 Doors and Gates. Doors and gates shall be low energy power operated doors or gates complying with Section 404.3. Doors shall remain open for 20 seconds minimum. On lifts with one door or with doors on opposite ends, the end door clear opening width shall be 32 inches (815 mm) minimum. On lifts with one door on a narrow end and one door on a long side, the end door clear opening width shall be 36 inches (915 mm) minimum. Side door clear opening width shall be 42 inches (1065 mm) minimum. Where a door is provided on a long side and on a narrow end of a lift, the side door shall be located with either the strike side or the hinge side in the corner furthest from the door on the narrow end.

EXCEPTIONS:

1. Doors or gates shall be permitted to be of the self-closing, manual type, where that door or gate provides access to a narrow end of the platform that serves only one landing. This exception shall not apply to doors or gates with ramps.

2. Lifts serving two landings maximum and having doors or gates on adjacent sides shall be permitted to have self closing manual doors or gates provided that the side door or gate is located with the strike side furthest from the end door. This exception shall not apply to door or gates with ramps.

410.2.2 Ramps. Ramp widths shall not be less than the platform opening they serve.

410.3 Floor Surfaces. Floor surfaces of platform lifts shall comply with Section 302.

410.4 Platform to Runway Clearance. The clearance between the platform sill and the edge of any runway landing shall be $1^1/_4$ inch (32 mm) maximum.

410.5 Clear Floor Space. Clear floor space of platform lifts shall comply with Section 410.5.

410.5.1 Lifts with Single Door or Doors on Opposite Ends. Platform lifts with a single door or with doors on opposite ends shall provide a clear floor width of 36 inches (915 mm) minimum and a clear floor depth of 48 inches (1220 mm) minimum.

410.5.2 Lifts with Doors on Adjacent Sides. Platform lifts with doors on adjacent sides shall provide a clear floor width of 42 inches (1065 mm) minimum and a clear floor depth of 60 inches (1525 mm) minimum.

> **EXCEPTION:** In existing buildings, platform lifts with doors on adjacent sides shall be permitted to provide a clear floor width of 36 inches (915 mm) and a clear floor depth of 60 inches (1525 mm).

410.6 Operable Parts. Controls for platform lifts shall comply with Section 309.

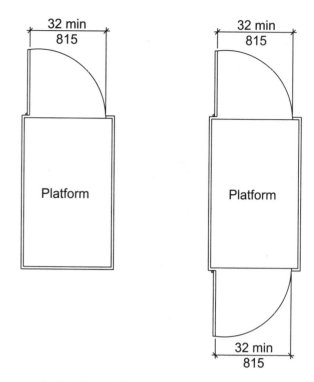

(a) Platform lift with door at one end or opposite ends

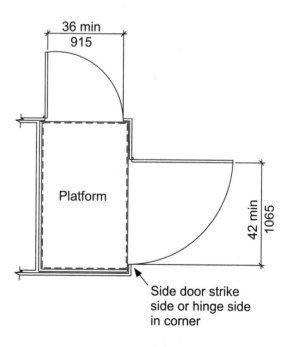

(b) Platform with doors on adjacent sides

FIG. 410.2.1
PLATFORM LIFT DOORS AND GATES

Chapter 5. General Site and Building Elements

501 General

501.1 Scope. General site and building elements required to be accessible by the scoping provisions adopted by the administrative authority shall comply with the applicable provisions of Chapter 5.

502 Parking Spaces

502.1 General. Accessible car and van parking spaces shall comply with Section 502.

502.2 Vehicle Space Size. Car parking spaces shall be 96 inches (2440 mm) minimum in width. Van parking spaces shall be 132 inches (3350 mm) minimum in width.

> **EXCEPTION:** Van parking spaces shall be permitted to be 96 inches (2440 mm) minimum in width where

the adjacent access aisle is 96 inches (2440 mm) minimum in width.

502.3 Vehicle Space Marking. Car and van parking spaces shall be marked to define the width. Where parking spaces are marked with lines, the width measurements of parking spaces and adjacent access aisles shall be made from the centerline of the markings.

> **EXCEPTION:** Where parking spaces or access aisles are not adjacent to another parking space or access aisle, measurements shall be permitted to include the full width of the line defining the parking space or access aisle.

502.4 Access Aisle. Car and van parking spaces shall have an adjacent access aisle complying with Section 502.4.

502.4.1 Location. Access aisles shall adjoin an accessible route. Two parking spaces shall be permitted to share a common access aisle. Access aisles shall not overlap with the vehicular way. Parking spaces shall be permitted to have access aisles placed on either side of the car or van parking space. Van parking spaces that are angled shall have access aisles located on the passenger side of the parking space.

502.4.2 Width. Access aisles serving car and van parking spaces shall be 60 inches (1525 mm) minimum in width.

502.4.3 Length. Access aisles shall extend the full length of the parking spaces they serve.

502.4.4 Marking. Access aisles shall be marked so as to discourage parking in them. Where access

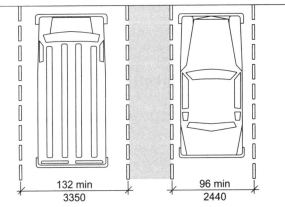

FIG. 502.2
VEHICLE PARKING SPACE SIZE

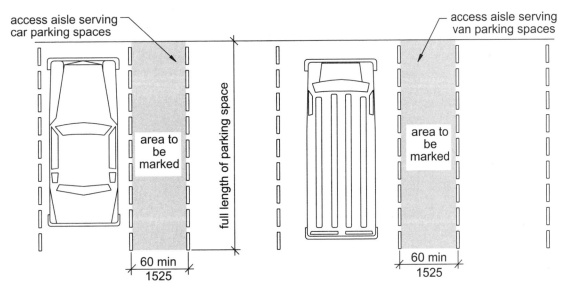

FIG. 502.4
PARKING SPACE ACCESS AISLE

aisles are marked with lines, the width measurements of access aisles and adjacent parking spaces shall be made from the centerline of the markings.

EXCEPTION: Where access aisles or parking spaces are not adjacent to another access aisle or parking space, measurements shall be permitted to include the full width of the line defining the access aisle or parking space.

502.5 Floor Surfaces. Parking spaces and access aisles shall comply with Section 302 and have surface slopes not steeper than 1:48. Access aisles shall be at the same level as the parking spaces they serve.

502.6 Vertical Clearance. A vertical clearance of 98 inches (2490 mm) minimum shall be provided at the following locations:

1. Parking spaces for vans.

2. The access aisles serving parking spaces for vans.

3. The vehicular routes serving parking spaces for vans.

502.7 Identification. Where accessible parking spaces are required to be identified by signs, the signs shall include the International Symbol of Accessibility complying with Section 703.6.3.1. Signs identifying van parking spaces shall contain the designation "van accessible." Such signs shall be 60 inches (1525 mm) minimum above the floor of the parking space, measured to the bottom of the sign.

502.8 Relationship to Accessible Routes. Parking spaces and access aisles shall be designed so that cars and vans, when parked, cannot obstruct the required clear width of adjacent accessible routes.

503 Passenger Loading Zones

503.1 General. Accessible passenger loading zones shall comply with Section 503.

503.2 Vehicle Pull-up Space Size. Passenger loading zones shall provide a vehicular pull-up space 96 inches (2440 mm) minimum in width and 20 feet (6095 mm) minimum in length.

503.3 Access Aisle. Passenger loading zones shall have an adjacent access aisle complying with Section 503.3.

503.3.1 Location. Access aisles shall adjoin an accessible route. Access aisles shall not overlap the vehicular way.

503.3.2 Width. Access aisles serving vehicle pull-up spaces shall be 60 inches (1525 mm) minimum in width.

503.3.3 Length. Access aisles shall be 20 feet (6095 mm) minimum in length.

503.3.4 Marking. Access aisles shall be marked so as to discourage parking in them.

503.4 Floor Surfaces. Vehicle pull-up spaces and access aisles serving them shall comply with Section 302 and shall have slopes not steeper than 1:48. Access aisles shall be at the same level as the vehicle pull-up space they serve.

503.5 Vertical Clearance. A vertical clearance of 114 inches (2895 mm) minimum shall be provided at the following locations:

1. Vehicle pull-up spaces;

2. The access aisles serving vehicle pull-up spaces;

3. A vehicular route from an entrance to the passenger loading zone, and;

4. A vehicular route from the passenger loading zone to a vehicular exit serving vehicle pull-up spaces.

504 Stairways

504.1 General. Accessible stairs shall comply with Section 504.

504.2 Treads and Risers. All steps on a flight of stairs shall have uniform riser height and uniform tread depth. Risers shall be 4 inches (100 mm) minimum and 7 inches (180 mm) maximum in height. Treads shall be 11 inches (280 mm) minimum in depth.

504.3 Open Risers. Open risers shall not be permitted.

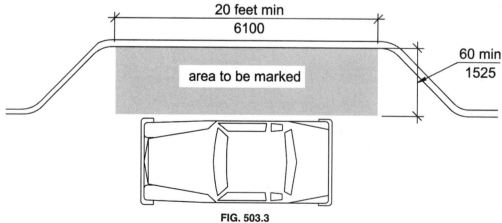

FIG. 503.3
PASSENGER LOADING ZONE ACCESS AISLE

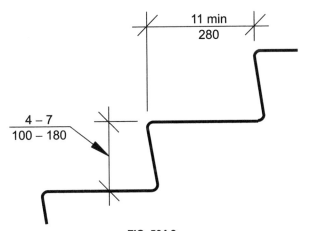

FIG. 504.2
TREADS AND RISERS FOR ACCESSIBLE STAIRWAYS

504.4 Tread Surface. Stair treads shall comply with Section 302 and shall have a slope not steeper than 1:48.

504.5 Nosings. The radius of curvature at the leading edge of the tread shall be $^1/_2$ inch (13 mm) maximum. Nosings that project beyond risers shall have the underside of the leading edge curved or beveled. Risers shall be permitted to slope under the tread at an angle of 30 degrees maximum from vertical. The permitted projection of the nosing shall be $1^1/_2$ inches (38 mm) maximum over the tread or floor below.

504.5.1 Visual contrast. The leading 2 inches (51 mm) of the tread shall have visual contrast of dark-on-light or light-on-dark from the remainder of the tread.

504.6 Handrails. Stairs shall have handrails complying with Section 505.

504.7 Wet Conditions. Stair treads and landings subject to wet conditions shall be designed to prevent the accumulation of water.

504.8 Lighting. Lighting for interior stairways shall comply with Section 504.8.

504.8.1 Illumination Level. Lighting facilities shall be capable of providing 10 foot-candles (108 lux) of illuminance measured at the center of tread surfaces and on landing surfaces within 24 inches (610 mm) of step nosings.

504.8.2 Lighting Controls. If provided, occupancy-sensing automatic controls shall activate the stairway lighting so the illuminance level required by Section 504.8.1 is provided on the entrance landing, each stair flight adjacent to the entrance landing, and on the landings above and below the entrance landing prior to any step being used.

504.9 Stair Level Identification. Stair level identification signs in raised characters and braille complying with Sections 703.3 and 703.4 shall be located at each floor level landing in all enclosed stairways adjacent to the door leading from the stairwell into the corridor to identify the floor level. The exit door discharging to the outside or to the level of exit discharge shall have a sign with raised characters and braille stating "EXIT."

505 Handrails

505.1 General. Handrails required by Section 405.8 for ramps, or Section 504.6 for stairs, shall comply with Section 505.

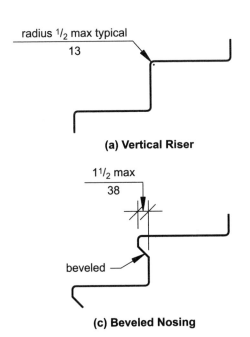

(a) Vertical Riser

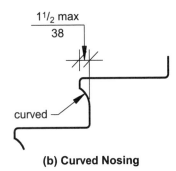

(b) Curved Nosing

(c) Beveled Nosing

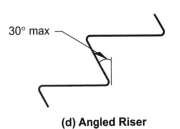

(d) Angled Riser

FIG. 504.5
STAIR NOSINGS

505.2 Location. Handrails shall be provided on both sides of stairs and ramps.

EXCEPTIONS:

1. In assembly seating areas, handrails shall not be required on both sides along aisle stairs, provided with a handrail either at the side or within the aisle.

2. In assembly seating areas, handrails shall not be required on the sides of ramped aisles serving seats.

505.3 Continuity. Handrails shall be continuous within the full length of each stair flight or ramp run. Inside handrails on switchback or dogleg stairs or ramps shall be continuous between flights or runs. Other handrails shall comply with Sections 505.10 and 307.

EXCEPTION: Handrails shall not be required to be continuous in aisles serving seating where handrails are discontinuous to provide access to seating and to permit crossovers within the aisles.

505.4 Height. Top of gripping surfaces of handrails shall be 34 inches (865 mm) minimum and 38 inches (965 mm) maximum vertically above stair nosings, ramp surfaces and walking surfaces. Handrails shall be at a consistent height above stair nosings, ramp surfaces and walking surfaces.

505.5 Clearance. Clearance between handrail gripping surface and adjacent surfaces shall be $1\frac{1}{2}$ inches (38 mm) minimum.

505.6 Gripping Surface. Gripping surfaces shall be continuous, without interruption by newel posts, other construction elements, or obstructions.

EXCEPTIONS:

1. Handrail brackets or balusters attached to the bottom surface of the handrail shall not be considered obstructions, provided the brackets or balusters comply with the following criteria:

    a. Not more than 20 percent of the handrail length is obstructed,

    b. Horizontal projections beyond the sides of the handrail occur $1\frac{1}{2}$ inches (38 mm) minimum below the bottom of the handrail, and provided that for each $\frac{1}{2}$ inch (13 mm) of additional handrail perimeter dimension above 4 inches (100 mm), the vertical clearance dimension of $1\frac{1}{2}$ inch (38 mm) can be reduced by $\frac{1}{8}$ inch (3.2 mm), and

    c. Edges shall be rounded.

2. Where handrails are provided along walking surfaces with slopes not steeper than 1:20, the bottoms of handrail gripping surfaces shall be permitted to be obstructed along their entire length where they are integral to crash rails or bumper guards.

505.7 Cross Section. Handrails shall have a cross section complying with Section 505.7.1 or 505.7.2.

505.7.1 Circular Cross Section. Handrails with a circular cross section shall have an outside diameter of $1\frac{1}{4}$ inches (32 mm) minimum and 2 inches (51 mm) maximum.

505.7.2 Noncircular Cross Sections. Handrails with a noncircular cross section shall have a perimeter dimension of 4 inches (100 mm) minimum and $6\frac{1}{4}$ inches (160 mm) maximum, and a cross-section dimension of $2\frac{1}{4}$ inches (57 mm) maximum.

505.8 Surfaces. Handrails, and any wall or other surfaces adjacent to them, shall be free of any sharp or abrasive elements. Edges shall be rounded.

505.9 Fittings. Handrails shall not rotate within their fittings.

505.10 Handrail Extensions. Handrails shall extend beyond and in the same direction of stair flights and ramp runs in accordance with Section 505.10.

EXCEPTIONS:

1. Continuous handrails at the inside turn of stairs and ramps.

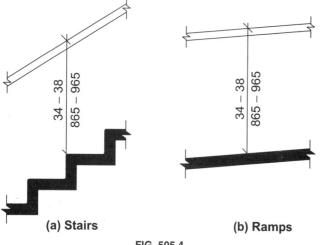

(a) Stairs   **(b) Ramps**

**FIG. 505.4
HANDRAIL HEIGHT**

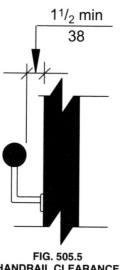

$1\frac{1}{2}$ min
38

**FIG. 505.5
HANDRAIL CLEARANCE**

2. Handrail extensions are not required in aisles serving seating where the handrails are discontinuous to provide access to seating and to permit crossovers within the aisle.

3. In alterations, full extensions of handrails shall not be required where such extensions would be hazardous due to plan configuration.

505.10.1 Top and Bottom Extension at Ramps. Ramp handrails shall extend horizontally above the landing 12 inches (305 mm) minimum beyond the top and bottom of ramp runs. Extensions shall return to a wall, guard, or floor, or shall be continuous to the handrail of an adjacent ramp run.

505.10.2 Top Extension at Stairs. At the top of a stair flight, handrails shall extend horizontally above the landing for 12 inches (305 mm) minimum beginning directly above the landing nosing. Extensions shall return to a wall, guard, or the landing surface, or shall be continuous to the handrail of an adjacent stair flight.

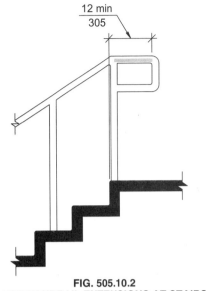

FIG. 505.10.2
TOP HANDRAIL EXTENSIONS AT STAIRS

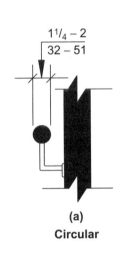

(a)
Circular

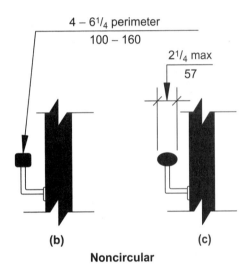

(b)   (c)
Noncircular

FIG. 505.7
HANDRAIL CROSS SECTION

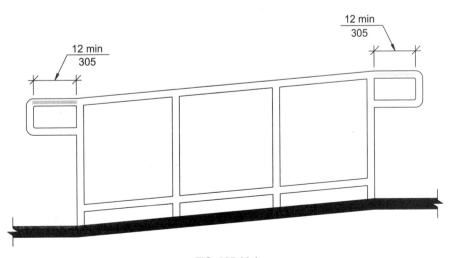

FIG. 505.10.1
TOP AND BOTTOM HANDRAIL EXTENSIONS AT RAMPS

505.10.3 Bottom Extension at Stairs. At the bottom of a stair flight, handrails shall extend at the slope of the stair flight for a horizontal distance equal to one tread depth beyond the bottom tread nosing. Extensions shall return to a wall, guard, or the landing surface, or shall be continuous to the handrail of an adjacent stair flight.

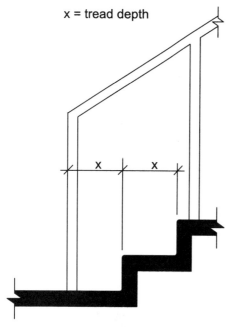

x = tread depth

FIG. 505.10.3
BOTTOM HANDRAIL EXTENSIONS AT STAIRS

506 Windows

506.1 General. Accessible windows shall have operable parts complying with Section 309.

Chapter 6. Plumbing Elements and Facilities

601 General

601.1 Scope. Plumbing elements and facilities required to be accessible by scoping provisions adopted by the administrative authority shall comply with the applicable provisions of Chapter 6.

602 Drinking Fountains

602.1 General. Accessible drinking fountains shall comply with Sections 602 and 307.

602.2 Clear Floor Space. A clear floor space complying with Section 305, positioned for a forward approach to the drinking fountain, shall be provided. Knee and toe space complying with Section 306 shall be provided. The clear floor space shall be centered on the drinking fountain.

EXCEPTIONS:

1. Drinking fountains for standing persons.

2. Drinking fountains primarily for children's use shall be permitted where the spout outlet is 30 inches (760 mm) maximum above the floor, a parallel approach complying with Section 305 is provided and the clear floor space is centered on the drinking fountain.

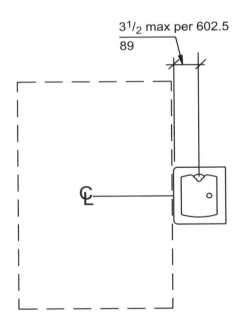

602.2
PARALLEL APPROACH AT DRINKING FOUNTAINS
PRIMARILY FOR CHILDREN'S USE—(EXCEPTION 2)

602.3 Operable Parts. Operable parts shall comply with Section 309.

602.4 Spout Outlet Height. Spout outlets of wheelchair accessible drinking fountains shall be 36 inches (915 mm) maximum above the floor. Spout outlets of drinking fountains for standing persons shall be 38 inches (965 mm) minimum and 43 inches (1090 mm) maximum above the floor.

602.5 Spout Location. The spout shall be located 15 inches (380 mm) minimum from the vertical support and 5 inches (125mm) maximum from the front edge of the drinking fountain, including bumpers. Where only a parallel approach is provided, the spout shall be located $3\frac{1}{2}$ inches (90 mm) maximum from the front edge of the drinking fountain, including bumpers.

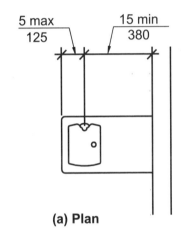

(a) Plan

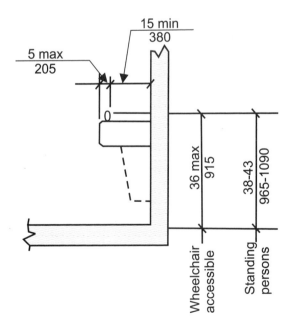

(b) Elevation

FIG. 602.5
DRINKING FOUNTAIN SPOUT LOCATION

602.6 Water Flow. The spout shall provide a flow of water 4 inches (100 mm) minimum in height. The angle of the water stream from spouts within 3 inches (75 mm) of the front of the drinking fountain shall be 30 degrees maximum, and from spouts between 3 inches (75 mm) and 5 inches (125 mm) from the front of the drinking fountain shall be 15 degrees maximum, measured horizontally relative to the front face of the drinking fountain.

603 Toilet and Bathing Rooms

603.1 General. Accessible toilet and bathing rooms shall comply with Section 603.

603.2 Clearances.

603.2.1 Turning Space. A turning space complying with Section 304 shall be provided within the room. The required turning space shall not be provided within a toilet compartment.

603.2.2 Door Swing. Doors shall not swing into the clear floor space or clearance for any fixture.

EXCEPTIONS:

1. Doors to a toilet or bathing room for a single occupant, accessed only through a private office and not for common use or public use shall be permitted to swing into the clear floor space, provided the swing of the door can be reversed to comply with Section 603.2.2.

2. Where the room is for individual use and a clear floor space complying with Section 305.3 is provided within the room beyond the arc of the door swing, the door shall not be required to comply with Section 603.2.2.

603.3 Mirrors. Where mirrors are located above lavatories, a mirror shall be located over the accessible lavatory and shall be mounted with the bottom edge of the reflecting surface 40 inches (1015 mm) maximum above the floor. Where mirrors are located above counters that do not contain lavatories, the mirror shall be mounted with the bottom edge of the reflecting surface 40 inches (1015 mm) maximum above the floor.

EXCEPTION: Other than within Accessible dwelling or sleeping units, mirrors are not required over the lavatories or counters if a mirror is located within the same toilet or bathing room and mounted with the bottom edge of the reflecting surface 35 inches (890 mm) maximum above the floor.

603.4 Coat Hooks and Shelves. Coat hooks shall be located within one of the reach ranges specified in Section 308. Shelves shall be 40 inches (1015 mm) minimum and 48 inches (1220 mm) maximum above the floor.

603.5 Diaper Changing Tables. Diaper changing tables shall comply with Sections 309 and 902

603.6 Operable Parts. Operable parts on towel dispensers and hand dryers serving accessible lavatories shall comply with Table 603.6.

604 Water Closets and Toilet Compartments

604.1 General. Accessible water closets and toilet compartments shall comply with Section 604. Compartments containing more than one plumbing fixture shall comply with Section 603. Wheelchair accessible compartments shall comply with Section 604.9. Ambulatory accessible compartments shall comply with Section 604.10.

EXCEPTION: Water closets and toilet compartments primarily for children's use shall be permitted to comply with Section 604.11 as applicable.

604.2 Location. The water closet shall be located with a wall or partition to the rear and to one side. The centerline of the water closet shall be 16 inches (405 mm) minimum and 18 inches (455 mm) maximum from the side wall or partition. Water closets located in ambulatory accessible compartments specified in Section 604.10 shall have the centerline of the water closet 17 inches (430 mm) minimum and 19 inches (485 mm) maximum from the side wall or partition.

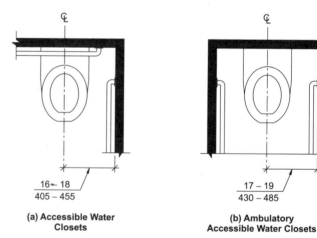

| 16 – 18 |
| 405 – 455 |

(a) Accessible Water Closets

| 17 – 19 |
| 430 – 485 |

(b) Ambulatory Accessible Water Closets

FIG. 604.2
WATER CLOSET LOCATION

TABLE 603.6
MAXIMUM REACH DEPTH AND HEIGHT

Maximum Reach Depth	0.5 inch (13 mm)	2 inches (51 mm)	5 inches (125 mm)	6 inches (150 mm)	9 inches (230 mm)	11 inches (280 mm)
Maximum Reach Height	48 inches (1220 mm)	46 inches (1170 mm)	42 inches (1065 mm)	40 inches (1015 mm)	36 inches (915 mm)	34 inches (865 mm)

604.3 Clearance.

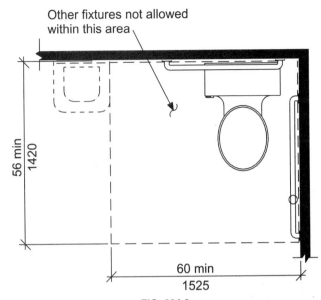

FIG. 604.3
SIZE OF CLEARANCE FOR WATER CLOSET

604.3.1 Clearance width. Clearance around a water closet shall be 60 inches (1525 mm) minimum in width, measured perpendicular from the sidewall.

604.3.2 Clearance Depth. Clearance around the water closet shall be 56 inches (1420 mm) minimum in depth, measured perpendicular from the rear wall.

604.3.3 Clearance Overlap. The required clearance around the water closet shall be permitted to overlap the water closet, associated grab bars, paper dispensers, sanitary napkin receptacles, coat hooks, shelves, accessible routes, clear floor space at other fixtures and the turning space. No other fixtures or obstructions shall be within the required water closet clearance.

604.4 Height. The height of water closet seats shall be 17 inches (430 mm) minimum and 19 inches (485 mm) maximum above the floor, measured to the top of the seat. Seats shall not be sprung to return to a lifted position.

EXCEPTION: A water closet in a toilet room for a single occupant, accessed only through a private office and not for common use or public use, shall not be required to comply with Section 604.4.

604.5 Grab Bars. Grab bars for water closets shall comply with Section 609 and shall be provided in accordance with Sections 604.5.1 and 604.5.2. Grab bars shall be provided on the rear wall and on the side wall closest to the water closet.

EXCEPTIONS:

1. Grab bars are not required to be installed in a toilet room for a single occupant, accessed only through a private office and not for common use or public use, provided reinforcement

has been installed in walls and located so as to permit the installation of grab bars complying with Section 604.5.

2. In detention or correction facilities, grab bars are not required to be installed in housing or holding cells or rooms that are specially designed without protrusions for purposes of suicide prevention.

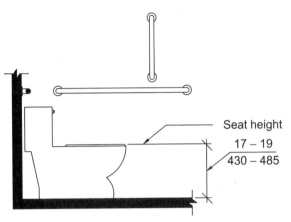

Note: For children's dimensions see Fig. 604.11.4

FIG. 604.4
WATER CLOSET SEAT HEIGHT

604.5.1 Fixed Side Wall Grab Bars. Fixed side-wall grab bars shall be 42 inches (1065 mm) minimum in length, located 12 inches (305 mm) maximum from the rear wall and extending 54 inches (1370 mm) minimum from the rear wall. In addition, a vertical grab bar 18 inches (455 mm) minimum in length shall be mounted with the bottom of the bar located 39 inches (990 mm) minimum and 41 inches (1040 mm)

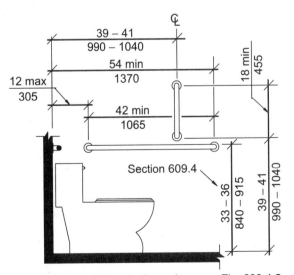

Note: For children's dimensions see Fig. 609.4.2

FIG. 604.5.1
SIDE WALL GRAB BAR FOR WATER CLOSET

maximum above the floor, and with the center line of the bar located 39 inches (990 mm) minimum and 41 inches (1040 mm) maximum from the rear wall.

> **EXCEPTION:** The vertical grab bar at water closets primarily for children's use shall comply with Section 609.4.2.

604.5.2 Rear Wall Grab Bars. The rear wall grab bar shall be 36 inches (915 mm) minimum in length, and extend from the centerline of the water closet 12 inches (305 mm) minimum on the side closest to the wall, and 24 inches (610 mm) minimum on the transfer side.

EXCEPTIONS:

1. The rear grab bar shall be permitted to be 24 inches (610 mm) minimum in length, centered on the water closet, where wall space does not permit a grab bar 36 inches (915 mm) minimum in length due to the location of a recessed fixture adjacent to the water closet.

2. Where an administrative authority requires flush controls for flush valves to be located in a position that conflicts with the location of the rear grab bar, that grab bar shall be permitted to be split or shifted to the open side of the toilet area.

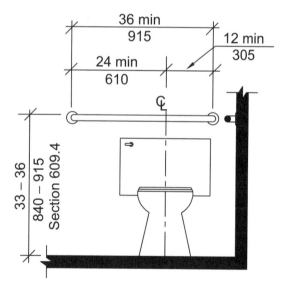

Note: For children's dimensions see Fig. 609.4.2

FIG. 604.5.2
REAR WALL GRAB BAR FOR WATER CLOSET

604.6 Flush Controls. Flush controls shall be hand operated or automatic. Hand operated flush controls shall comply with Section 309. Flush controls shall be located on the open side of the water closet.

> **EXCEPTION:** In ambulatory accessible compartments complying with Section 604.10, flush controls shall be permitted to be located on either side of the water closet.

604.7 Dispensers. Toilet paper dispensers shall comply with Section 309.4. Where the dispenser is located above the grab bar, the outlet of the dispenser shall be located within an area 24 inches (610 mm) minimum and 36 inches (915 mm) maximum from the rear wall. Where the dispenser is located below the grab bar, the outlet of the dispenser shall be located within an area 24 inches (610 mm) minimum and 42 inches (1065 mm) maximum from the rear wall. The outlet of the dispenser shall be located 18 inches (455 mm) minimum and 48 inches (1220 mm) maximum above the floor. Dispensers shall comply with Section 609.3. Dispensers shall not be of a type that control delivery, or do not allow continuous paper flow.

604.8 Coat Hooks and Shelves. Coat hooks provided within toilet compartments shall be 48 inches (1220 mm) maximum above the floor. Shelves shall be 40 inches (1015 mm) minimum and 48 inches (1220 mm) maximum above the floor.

604.9 Wheelchair Accessible Compartments.

604.9.1 General. Wheelchair accessible compartments shall comply with Section 604.9.

604.9.2 Size. Toilet compartments shall comply with Section 604.9.2.1 or 604.9.2.2 as applicable.

604.9.2.1 Minimum area. The minimum area of a wheelchair accessible compartment shall be 60 inches (1525 mm) minimum in width measured perpendicular to the side wall, and 56 inches (1420 mm) minimum in depth for wall hung water closets, and 59 inches (1500 mm) minimum in depth for floor mounted water closets measured perpendicular to the rear wall.

604.9.2.2 Compartment for children's use. The minimum area of a wheelchair accessible compartment primarily for children's use shall be 60 inches (1525 mm) minimum in width measured perpendicular to the side wall, and 59 inches (1500 mm) minimum in depth for wall hung and floor mounted water closets measured perpendicular to the rear wall.

604.9.3 Doors. Toilet compartment doors, including door hardware, shall comply with Section 404, except if the approach is to the latch side of the compartment door clearance between the door side of the stall and any obstruction shall be 42 inches (1065 mm) minimum. The door shall be self-closing. A door pull complying with Section 404.2.6 shall be placed on both sides of the door near the latch. Toilet compartment doors shall not swing into the required minimum area of the compartment.

604.9.3.1 Door Opening Location. The farthest edge of toilet compartment door opening shall be located in the front wall or partition or in the side wall or partition as required by Table 604.9.3.1.

604.9.4 Approach. Wheelchair accessible compartments shall be arranged for left-hand or right-hand approach to the water closet.

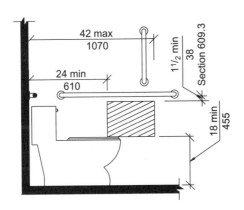

(a) Protruding Dispenser Below Grab Bar

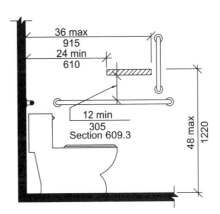

(b) Protruding Dispenser Above Grab Bar

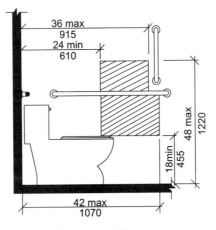

(c) Recessed Dispenser

Note: For children's dimensions see Fig. 604.11.7 dispenser outlet location

**FIG. 604.7
DISPENSER OUTLET LOCATION**

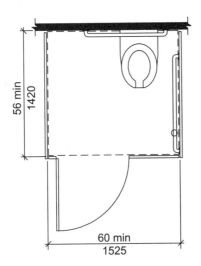

(a) Wall-Hung Water Closet – Adult

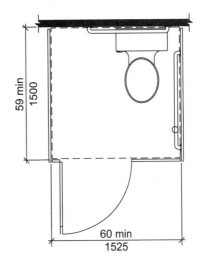

**(b) Floor-Mounted Water Closet – Adult
Wall-Hung and
Floor-Mounted Water Closet – Children**

**FIG. 604.9.2
WHEELCHAIR ACCESSIBLE TOILET COMPARTMENTS**

TABLE 604.9.3.1—DOOR OPENING LOCATION

Door Opening Location	Measured From	Dimension
Front Wall or Partition	From the side wall or partition closest to the water closet	56 inches (1420 mm) minimum
	or	
	From the side wall or partition farthest from the water closet	4 inches (100 mm) maximum
Side Wall or Partition Wall-Hung Water Closet	From the rear wall	52 inches (1320 mm) minimum
	or	
	From the front wall or partition	4 inches (100 mm) maximum
Side Wall or Partition Floor-Mounted Water Closet	From the rear wall	55 inches (1395 mm) minimum
	or	
	From the front wall or partition	4 inches (100 mm) maximum

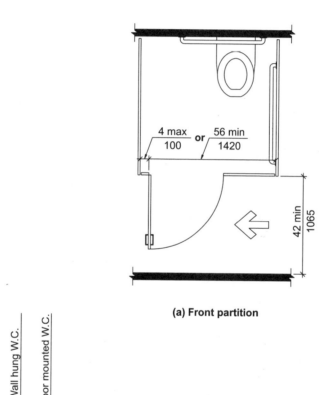

(a) Front partition

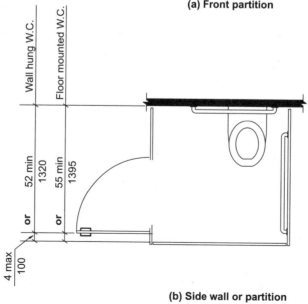

(b) Side wall or partition

FIG. 604.9.3.1
WHEELCHAIR ACCESSIBLE COMPARTMENT DOOR OPENINGS

604.9.5 Toe Clearance. Toe clearance for compartments primarily for children's use shall comply with Section 604.9.5.2. Toe clearance for other wheelchair accessible compartments shall comply with Section 604.9.5.1.

604.9.5.1 Toe Clearance at Compartments. The front partition and at least one side partition shall provide a toe clearance of 9 inches (230 mm) minimum above the floor and extending 6 inches (150 mm) beyond the compartment side face of the partition, exclusive of partition support members.

EXCEPTIONS:

1. Toe clearance at the front partition is not required in a compartment greater than 62 inches (1575 mm) in depth with a wall-hung water closet, or greater than 65 inches (1650 mm) in depth with a floor-mounted water closet.

2. Toe clearance at the side partition is not required in a compartment greater than 66 inches (1675 mm) in width.

604.9.5.2 Toe Clearance at Compartments for Children's Use. The front partition and at least one side partition of compartments primarily for children's use shall provide a toe clearance of 12 inches (305 mm) minimum above the floor and extending 6 inches (150 mm) beyond the compart-

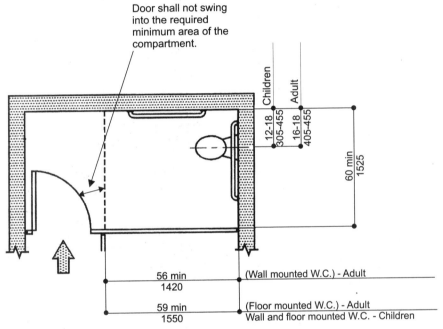

FIG. 604.9.3.1 (C)
WHEELCHAIR ACCESSIBLE COMPARTMENT DOOR OPENINGS—ALTERNATE

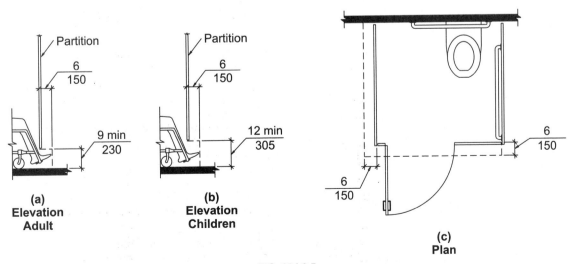

FIG. 604.9.5
WHEELCHAIR ACCESSIBLE COMPARTMENT TOE CLEARANCE

ment side face of the partition, exclusive of partition support members.

EXCEPTIONS:

1. Toe clearance at the front partition is not required in a compartment greater than 65 inches (1650 mm) in depth.

2. Toe clearance at the side partition is not required in a compartment greater than 66 inches (1675 mm) in width.

604.9.6 Grab Bars. Grab bars shall comply with Section 609. Side wall grab bars complying with Section 604.5.1 located on the wall closest to the water closet, and a rear wall grab bar complying with Section 604.5.2, shall be provided.

604.10 Ambulatory Accessible Compartments.

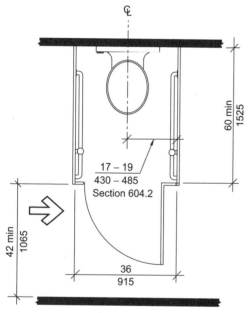

FIG. 604.10
AMBULATORY ACCESSIBLE COMPARTMENT

604.10.1 General. Ambulatory accessible compartments shall comply with Section 604.10.

604.10.2 Size. The minimum area of an ambulatory accessible compartment shall be 60 inches (1525 mm) minimum in depth and 36 inches (915 mm) in width.

604.10.3 Doors. Toilet compartment doors, including door hardware, shall comply with Section 404, except if the approach is to the latch side of the compartment door the clearance between the door side of the compartment and any obstruction shall be 42 inches (1065 mm) minimum. The door shall be self-closing. A door pull complying with Section 404.2.6 shall be placed on both sides of the door near the latch. Compartment doors shall not swing into the required minimum area of the compartment.

604.10.4 Grab Bars. Grab bars shall comply with Section 609. Side wall grab bars complying with Sec-

tion 604.5.1 shall be provided on both sides of the compartment.

604.11 Water Closets and Toilet Compartments for Children's Use.

604.11.1 General. Accessible water closets and toilet compartments primarily for children's use shall comply with Section 604.11.

604.11.2 Location. The water closet primarily for children's use shall be located with a wall or partition to the rear and to one side. The centerline of the water closet shall be 12 inches (305 mm) minimum and 18 inches (455 mm) maximum from the side wall or partition. Water closets located in ambulatory accessible toilet compartments specified in Section 604.10 shall be located as specified in Section 604.2.

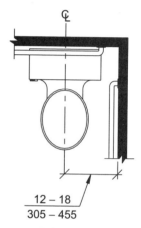

FIG. 604.11.2
CHILDREN'S WATER CLOSET LOCATION

604.11.3 Clearance. A clearance around the water closet primarily for children's use complying with Section 604.3 shall be provided.

604.11.4 Height. The height of water closet seats primarily for children's use shall be 11 inches (280 mm) minimum and 17 inches (430 mm) maximum above the floor, measured to the top of the seat. Seats shall not be sprung to return to a lifted position.

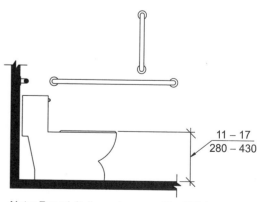

Note: For adult dimensions see Fig. 604.4

FIG. 604.11.4
CHILDREN'S WATER CLOSET HEIGHT

604.11.5 Grab Bars. Grab bars for water closets primarily for children's use shall comply with Section 604.5.

604.11.6 Flush Controls. Flush controls primarily for children's use shall be hand operated or automatic. Hand operated flush controls shall comply with Sections 309.2 and 309.4 and shall be installed 36 inches (915 mm) maximum above the floor. Flush controls shall be located on the open side of the water closet.

> **EXCEPTION:** In ambulatory accessible compartments complying with Section 604.10, flush controls shall be permitted to be located on either side of the water closet.

604.11.7 Dispensers. Toilet paper dispensers primarily for children's use shall comply with Section 309.4. The outlet of dispensers shall be located within an area 24 inches (610 mm) minimum and 42 inches (1065 mm) maximum from the rear wall. The outlet of the dispenser shall be 14 inches (355 mm) minimum and 19 inches (485 mm) maximum above the floor. There shall be a clearance of $1^1/_2$ inches (38 mm) minimum below the grab bar. Dispensers shall not be of a type that control delivery or do not allow continuous paper flow.

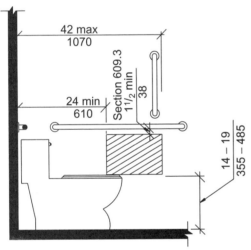

Note: For adult dimensions see Fig. 604.7

FIG. 604.11.7
CHILDREN'S DISPENSER OUTLET LOCATION

604.11.8 Toilet Compartments. Toilet compartments primarily for children's use shall comply with Sections 604.9 and 604.10, as applicable.

605 Urinals

605.1 General. Accessible urinals shall comply with Section 605.

605.2 Height and Depth. Urinals shall be of the stall type or shall be of the wall hung type with the rim at 17 inches (430 mm) maximum above the floor. Wall hung urinals shall be $13^1/_2$ inches (345 mm) minimum in

depth measured from the outer face of the urinal rim to the wall.

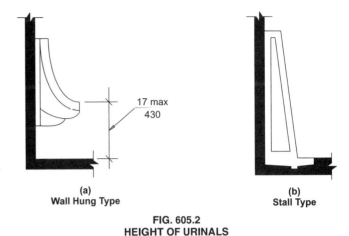

FIG. 605.2
HEIGHT OF URINALS

605.3 Clear Floor Space. A clear floor space complying with Section 305, positioned for forward approach, shall be provided.

605.4 Flush Controls. Flush controls shall be hand operated or automatic. Hand operated flush controls shall comply with Section 309.

606 Lavatories and Sinks

606.1 General. Accessible lavatories and sinks shall comply with Section 606.

606.2 Clear Floor Space. A clear floor space complying with Section 305.3, positioned for forward approach, shall be provided. Knee and toe clearance complying with Section 306 shall be provided. The dip of the overflow shall not be considered in determining knee and toe clearances.

> **EXCEPTIONS:**
>
> 1. A parallel approach complying with Section 305 and centered on the sink, shall be permitted to a kitchen sink in a space where a cook top or conventional range is not provided.
>
> 2. The requirement for knee and toe clearance shall not apply to a lavatory in a toilet or bathing facility for a single occupant, accessed only through a private office and not for common use or public use.
>
> 3. A knee clearance of 24 inches (610 mm) minimum above the floor shall be permitted at lavatories and sinks used primarily by children ages 6 through 12 where the rim or counter surface is 31 inches (785 mm) maximum above the floor.
>
> 4. A parallel approach complying with Section 305 and centered on the sink, shall be permitted at lavatories and sinks used primarily by children ages 5 and younger.

5. The requirement for knee and toe clearance shall not apply to more than one bowl of a multibowl sink.

6. A parallel approach complying with Section 305 and centered on the sink, shall be permitted at wet bars.

606.3 Height. The front of lavatories and sinks shall be 34 inches (865 mm) maximum above the floor, measured to the higher of the rim or counter surface.

EXCEPTION: A lavatory in a toilet or bathing facility for a single occupant, accessed only through a private office and not for common use or public use, shall not be required to comply with Section 606.3.

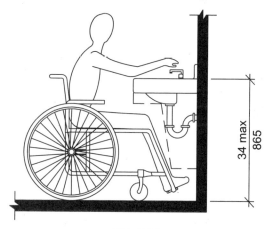

**FIG. 606.3
HEIGHT OF LAVATORIES AND SINKS**

606.4 Faucets. Faucets shall comply with Section 309. Hand-operated metering faucets shall remain open for 10 seconds minimum.

606.5 Lavatories with Enhanced Reach Range. Where enhanced reach range is required at lavatories, faucets and soap dispenser controls shall have a reach depth of 11 inches (280 mm) maximum or, if automatic, shall be activated within a reach depth of 11 inches (280 mm) maximum. Water and soap flow shall be provided with a reach depth of 11 inches (280 mm) maximum.

606.6 Exposed Pipes and Surfaces. Water supply and drainpipes under lavatories and sinks shall be insulated or otherwise configured to protect against contact. There shall be no sharp or abrasive surfaces under lavatories and sinks.

607 Bathtubs

607.1 General. Accessible bathtubs shall comply with Section 607.

607.2 Clearance. A clearance in front of bathtubs extending the length of the bathtub and 30 inches (760 mm) minimum in depth shall be provided. Where a permanent seat is provided at the head end of the bathtub, the clearance shall extend 12 inches (305 mm) minimum beyond the wall at the head end of the bathtub.

607.3 Seat. A permanent seat at the head end of the bathtub or a removable in-tub seat shall be provided. Seats shall comply with Section 610.

607.4 Grab Bars. Grab bars shall comply with Section 609 and shall be provided in accordance with Section 607.4.1 or 607.4.2.

EXCEPTION: Grab bars shall not be required to be installed in a bathing facility for a single occupant accessed only through a private office and not for common use or public use, provided reinforcement has been installed in walls and located so as to permit the installation of grab bars complying with Section 607.4.

607.4.1 Bathtubs with Permanent Seats. For bathtubs with permanent seats, grab bars complying with Section 607.4.1 shall be provided.

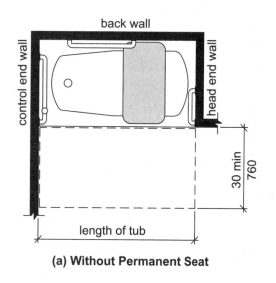

(a) Without Permanent Seat

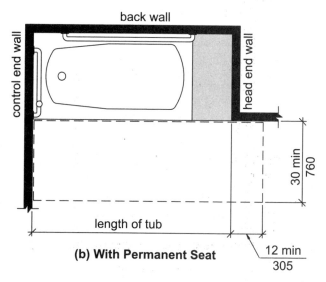

(b) With Permanent Seat

**FIG. 607.2
CLEARANCE FOR BATHTUBS**

607.4.1.1 Back Wall. Two horizontal grab bars shall be provided on the back wall, one complying with Section 609.4 and the other located 8 inches (205 mm) minimum and 10 inches (255 mm) maximum above the rim of the bathtub. Each grab bar shall be located 15 inches (380 mm) maximum from the head end wall and extend to 12 inches (305 mm) maximum from the control end wall.

607.4.1.2 Control End Wall. Control end wall grab bars shall comply with Section 607.4.1.2.

EXCEPTION: An L-shaped continuous grab bar of equivalent dimensions and positioning shall be permitted to serve the function of separate vertical and horizontal grab bars.

607.4.1.2.1 Horizontal Grab Bar. A horizontal grab bar 24 inches (610 mm) minimum in length shall be provided on the control end wall beginning near the front edge of the bathtub and extending toward the inside corner of the bathtub.

607.4.1.2.2 Vertical Grab Bar. A vertical grab bar 18 inches (455 mm) minimum in length shall be provided on the control end wall 3 inches (75 mm) minimum and 6 inches (150 mm) maximum above the horizontal grab bar, and 4 inches (100 mm) maximum inward from the front edge of the bathtub.

607.4.2 Bathtubs without Permanent Seats. For bathtubs without permanent seats, grab bars complying with Section 607.4.2 shall be provided.

607.4.2.1 Back Wall. Two horizontal grab bars shall be provided on the back wall, one complying

with Section 609.4 and the other located 8 inches (205 mm) minimum and 10 inches (255 mm) maximum above the rim of the bathtub. Each grab bar shall be 24 inches (610 mm) minimum in length, located 24 inches (610 mm) maximum from the head end wall and extend to 12 inches (305 mm) maximum from the control end wall.

607.4.2.2 Control End Wall. Control end wall grab bars shall comply with Section 607.4.1.2.

607.4.2.3 Head End Wall. A horizontal grab bar 12 inches (305 mm) minimum in length shall be provided on the head end wall at the front edge of the bathtub.

607.5 Controls. Controls, other than drain stoppers, shall be provided on an end wall, located between the bathtub rim and grab bar, and between the open side of the bathtub and the centerline of the width of the bathtub. Controls shall comply with Section 309.4.

607.6 Hand Shower. A hand shower with a hose 59 inches (1500 mm) minimum in length, that can be used as both a fixed shower head and as a hand shower, shall be provided. The hand shower shall have a control with a nonpositive shut-off feature. Where provided, an adjustable-height hand shower mounted on a vertical bar shall be installed so as to not obstruct the use of grab bars.

607.7 Bathtub Enclosures. Enclosures for bathtubs shall not obstruct controls, faucets, shower and spray units or obstruct transfer from wheelchairs onto bathtub seats or into bathtubs. Enclosures on bathtubs shall not have tracks installed on the rim of the bathtub.

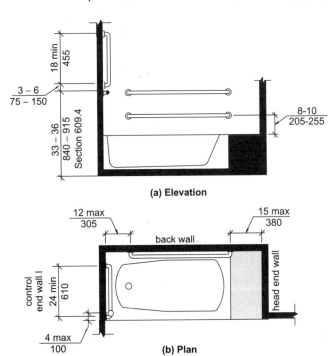

FIG. 607.4.1
GRAB BARS FOR BATHTUBS WITH PERMANENT SEATS

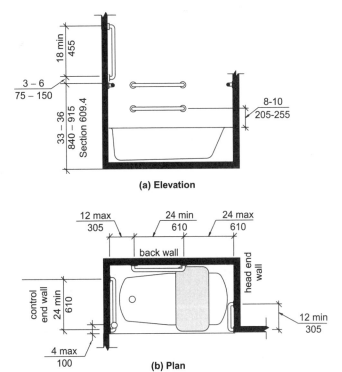

FIG. 607.4.2
GRAB BARS FOR BATHTUBS WITHOUT PERMANENT SEATS

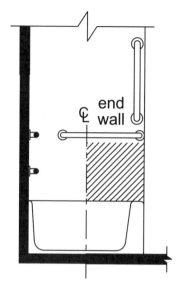

FIG. 607.5
LOCATION OF BATHTUB CONTROLS

607.8 Water Temperature. Bathtubs shall deliver water that is 120°F (49°C) maximum.

608 Shower Compartments

608.1 General. Accessible shower compartments shall comply with Section 608.

608.2 Size, clearance and seat. Shower compartments shall have sizes, clearances and seats complying with Section 608.2.

608.2.1 Transfer-type Shower Compartments. Transfer-type shower compartments shall comply with Section 608.2.1.

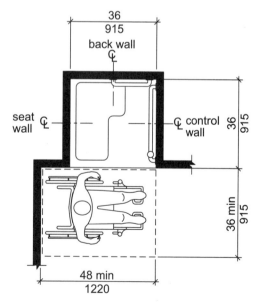

Note: inside finished dimensions measured at the center points of opposing sides

FIG. 608.2.1
TRANSFER-TYPE SHOWER
COMPARTMENT SIZE AND CLEARANCE

608.2.1.1 Size. Transfer-type shower compartments shall have a clear inside dimension of 36 inches (915 mm) in width and 36 inches (915 mm) in depth, measured at the center point of opposing sides. An entry 36 inches (915 mm) minimum in width shall be provided.

608.2.1.2 Clearance. A clearance of 48 inches (1220 mm) minimum in length measured perpendicular from the control wall, and 36 inches (915 mm) minimum in depth shall be provided adjacent to the open face of the compartment.

608.2.1.3 Seat. A folding or non-folding seat complying with Section 610 shall be provided on the wall opposite the control wall.

> **Exception:** A seat is not required to be installed in a shower for a single occupant, accessed only through a private office and not for common use or public use, provided reinforcement has been installed in walls and located so as to permit the installation of a shower seat.

608.2.2 Standard Roll-in-type Shower Compartments. Standard roll-in-type shower compartments shall comply with Section 608.2.2.

608.2.2.1 Size. Standard roll-in-type shower compartments shall have a clear inside dimension of 60 inches (1525 mm) minimum in width and 30 inches (760 mm) minimum in depth, measured at the center point of opposing sides. An entry 60 inches (1525 mm) minimum in width shall be provided.

608.2.2.2 Clearance. A clearance of 60 inches (1525 mm) minimum in length adjacent to the 60-inch (1525 mm) width of the open face of the shower compartment, and 30 inches (760 mm) minimum in depth, shall be provided.

> **EXCEPTION:** A lavatory complying with Section 606 shall be permitted at the end of the clearance opposite the seat.

608.2.2.3 Seat. A folding seat complying with Section 610 shall be provided on an end wall.

> **EXCEPTIONS:**
> 1. A seat is not required to be installed in a shower for a single occupant accessed only through a private office and not for common use or public use, provided reinforcement has been installed in walls and located so as to permit the installation of a shower seat.
> 2. A fixed seat shall be permitted where the seat does not overlap the minimum clear inside dimension required by Section 608.2.2.1.

608.2.3 Alternate Roll-in-type Shower Compartments. Alternate roll-in-type shower compartments shall comply with Section 608.2.3.

608.2.3.1 Size. Alternate roll-in shower compartments shall have a clear inside dimension of 60 inches (1525 mm) minimum in width, and 36 inches (915 mm) in depth, measured at the center point of opposing sides. An entry 36 inches (915 mm) minimum in width shall be provided at one end of the 60-inch (1525 mm) width of the compartment. A seat wall, 24 inches (610 mm) minimum and 36 inches (915 mm) maximum in length, shall be provided on the entry side of the compartment.

608.2.3.2 Seat. A folding seat complying with Section 610 shall be provided on the seat wall opposite the back wall.

EXCEPTION: A seat is not required to be installed in a shower for a single occupant, accessed only through a private office and not for common use or public use, provided reinforcement has been installed in walls and located so as to permit the installation of a shower seat.

608.3 Grab Bars. Grab bars shall comply with Section 609 and shall be provided in accordance with Section 608.3. Where multiple grab bars are used, required horizontal grab bars shall be installed at the same height above the floor.

EXCEPTION: Grab bars are not required to be installed in a shower for a single occupant, accessed only through a private office and not for common use or public use, provided reinforcement has been installed in walls and located so as to permit the installation of grab bars complying with Section 608.3.

608.3.1 Transfer-Type Showers. Grab bars for transfer type showers shall comply with Section 608.3.1.

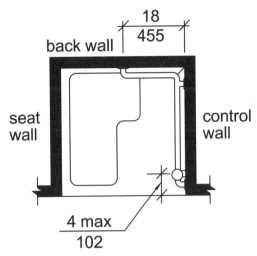

**FIG. 608.3.1
GRAB BARS IN TRANSFER-TYPE SHOWERS**

608.3.1.1 Horizontal Grab Bars. Horizontal grab bars shall be provided across the control wall and on the back wall to a point 18 inches (455 mm) from the control wall.

608.3.1.2 Vertical Grab Bar. A vertical grab bar 18 inches (455 mm) minimum in length shall be provided on the control end wall 3 inches (75 mm) minimum and 6 inches (150 mm) maximum above the horizontal grab bar, and 4 inches (100 mm) maximum inward from the front edge of the shower.

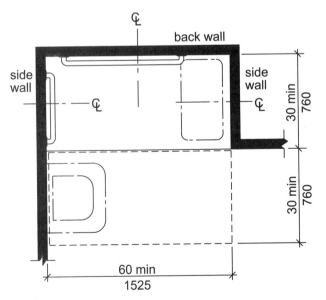

Note: inside finished dimensions measured at the center points of opposing sides

**FIG. 608.2.2
STANDARD ROLL-IN-TYPE SHOWER
COMPARTMENT SIZE AND CLEARANCE**

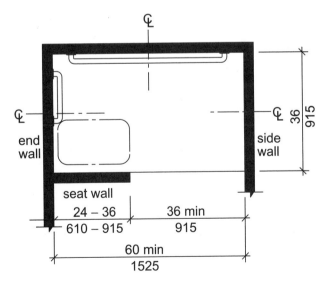

Note: inside finished dimensions measured at the center points of opposing sides

**FIG. 608.2.3
ALTERNATE ROLL-IN-TYPE SHOWER
COMPARTMENT SIZE AND CLEARANCE**

608.3.2 Standard Roll-in-Type Showers. In standard roll-in type showers, a grab bar shall be provided on the back wall beginning at the edge of the seat. The grab bars shall not be provided above the seat. The back wall grab bar shall extend the length of the wall but shall not be required to exceed 48 inches (1220 mm) in length. Where a side wall is provided opposite the seat within 72 inches (1830 mm) of the seat wall, a grab bar shall be provided on the side wall opposite the seat. The side wall grab bar shall extend the length of the wall but shall not be required to exceed 30 inches (760 mm) in length. Grab bars shall be 6 inches (150 mm) maximum from the adjacent wall.

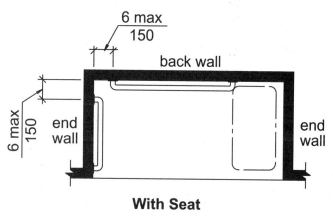

With Seat

FIG. 608.3.2
GRAB BARS IN STANDARD ROLL-IN-TYPE SHOWERS

608.3.3 Alternate Roll-in-Type Showers. In alternate roll-in type showers, grab bars shall be provided on the back wall and the end wall adjacent to the seat. Grab bars shall not be provided above the seat. Grab bars shall be 6 inches (150 mm) maximum from the adjacent wall.

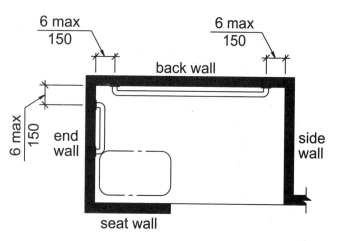

FIG. 608.3.3
GRAB BARS IN ALTERNATE ROLL-IN-TYPE SHOWERS

608.4 Controls and Hand Showers. Controls and hand showers shall comply with Sections 608.4 and 309.4.

608.4.1 Transfer-Type Showers. In transfer-type showers, the controls and hand shower shall be located:

1. On the control wall opposite the seat.

2. At a height of 38 inches (965 mm) minimum and 48 inches (1220 mm) maximum above the shower floor, and

3. 15 inches (380 mm) maximum, from the centerline of the control wall toward the shower opening.

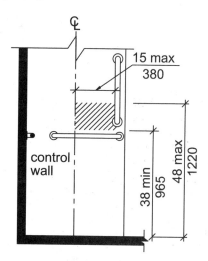

FIG. 608.4.1
TRANSFER-TYPE SHOWER
CONTROLS AND HANDSHOWER LOCATION

608.4.2 Standard Roll-in Showers. In standard roll-in showers, the controls and hand shower shall be located on the back wall above the grab bar, 48 inches (1220 mm) maximum above the shower floor and 16 inches (405 mm) minimum and 27 inches (685 mm) maximum from the end wall behind the seat.

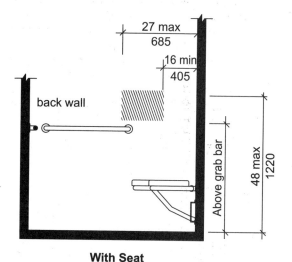

With Seat

FIG. 608.4.2
STANDARD ROLL-IN-TYPE SHOWER
CONTROL AND HANDSHOWER LOCATION

608.4.3 Alternate Roll-in Showers. In alternate roll-in showers, the controls and hand shower shall be located 38 inches (965 mm) minimum and 48 inches (1220 mm) maximum above the shower floor. In alternate roll-in showers with controls and hand shower located on the end wall adjacent to the seat, the controls and hand shower shall be 27 inches (685 mm) maximum from the seat wall. In alternate roll-in showers with the controls and hand shower located on the back wall opposite the seat, the controls and hand shower shall be located within 15 inches (380 mm), left or right, of the centerline of the seat.

608.5 Hand Showers. A hand shower with a hose 59 inches (1500 mm) minimum in length, that can be used both as a fixed shower head and as a hand shower, shall be provided. The hand shower shall have a control with a nonpositive shut-off feature. Where provided, an adjustable-height hand shower mounted on a vertical bar shall be installed so as to not obstruct the use of grab bars.

EXCEPTION: In other than Accessible units and Type A units, a fixed shower head located 48 inches (1220 mm) maximum above the shower floor shall be permitted in lieu of a hand shower.

608.6 Thresholds. Thresholds in roll-in-type shower compartments shall be $1/_2$ inch (13 mm) maximum in height in accordance with Section 303. In transfer-type shower compartments, thresholds $1/_2$ inch (13 mm) maximum in height shall be beveled, rounded, or vertical.

EXCEPTION: In existing facilities, in transfer-type shower compartments where provision of a threshold $1/_2$ inch (13 mm) in height would disturb the structural reinforcement of the floor slab, a threshold 2 inches (51 mm) maximum in height shall be permitted.

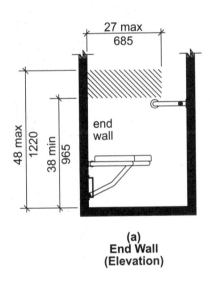

**(a)
End Wall
(Elevation)**

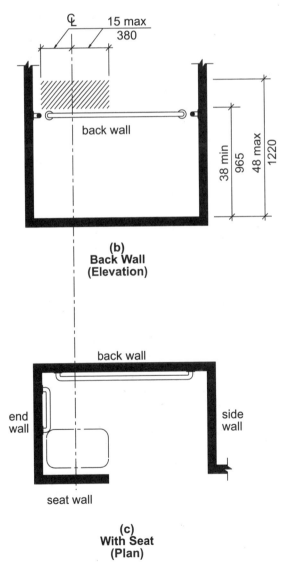

**(b)
Back Wall
(Elevation)**

**(c)
With Seat
(Plan)**

**FIG. 608.4.3
ALTERNATE ROLL-IN-TYPE SHOWER CONTROL AND HANDSHOWER LOCATION**

608.7 Shower Enclosures. Shower compartment enclosures for shower compartments shall not obstruct controls or obstruct transfer from wheelchairs onto shower seats.

608.8 Water Temperature. Showers shall deliver water that is 120°F (49°C) maximum.

609 Grab Bars

609.1 General. Grab bars in accessible toilet or bathing facilities shall comply with Section 609.

609.2 Cross Section. Grab bars shall have a cross section complying with Section 609.2.1 or 609.2.2.

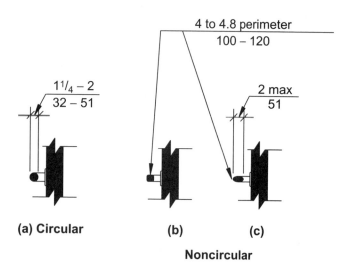

FIG. 609.2
SIZE OF GRAB BARS

609.2.1 Circular Cross Section. Grab bars with a circular cross section shall have an outside diameter of $1\frac{1}{4}$ inch (32 mm) minimum and 2 inches (51 mm) maximum.

609.2.2 Noncircular Cross Section. Grab bars with a noncircular cross section shall have a cross section dimension of 2 inches (51 mm) maximum, and a perimeter dimension of 4 inches (100 mm) minimum and 4.8 inches (120 mm) maximum.

609.3 Spacing. The space between the wall and the grab bar shall be $1\frac{1}{2}$ inches (38 mm). The space between the grab bar and projecting objects below and at the ends of the grab bar shall be $1\frac{1}{2}$ inches (38 mm) minimum. The space between the grab bar and projecting objects above the grab bar shall be 12 inches (305 mm) minimum.

EXCEPTIONS:

1. The space between the grab bars and shower controls, shower fittings, and other grab bars above the grab bar shall be permitted to be $1\frac{1}{2}$ inches (38 mm) minimum.

2. Recessed dispensers projecting from the wall $\frac{1}{4}$ inch (6.4 mm) maximum measured from the face of the dispenser and complying with Section

604.7 shall be permitted within the 12-inch (305 mm) space above and the $1\frac{1}{2}$ inch (38 mm) spaces below and at the ends of the grab bar.

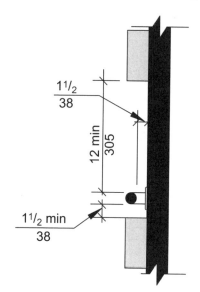

FIG. 609.3
SPACING OF GRAB BARS

609.4 Position of Grab Bars.

609.4.1 General. Grab bars shall be installed in a horizontal position, 33 inches (840 mm) minimum and 36 inches (915 mm) maximum above the floor measured to the top of the gripping surface or shall be installed as required by Items 1 through 3.

1. The lower grab bar on the back wall of a bathtub shall comply with Section 607.4.1.1 or 607.4.2.1.

2. Vertical grab bars shall comply with Sections 604.5.1, 607.4.1.2.2, 607.4.2.2, and 608.3.1.2.

3. Grab bars at water closets primarily for children's use shall comply with Section 609.4.2.

609.4.2 Position of Children's Grab Bars. At water closets primarily for children's use complying with Section 604.11, grab bars shall be installed in a horizontal position 18 inches (455 mm) minimum and 27 inches (685 mm) maximum above the floor measured to the top of the gripping surface. A vertical grab bar shall be mounted with the bottom of the bar located between 21 inches (535 mm) minimum and 30 inches (760 mm) maximum above the floor and with the centerline of the bar located between 34 inches (865 mm) minimum and 36 inches (915 mm) maximum from the rear wall.

609.5 Surface Hazards. Grab bars, and any wall or other surfaces adjacent to grab bars, shall be free of sharp or abrasive elements. Edges shall be rounded.

609.6 Fittings. Grab bars shall not rotate within their fittings.

609.7 Installation and Configuration. Grab bars shall be installed in any manner that provides a gripping surface at the locations specified in this standard and does not obstruct the clear floor space. Horizontal and vertical grab bars shall be permitted to be separate bars, a single piece bar, or combination thereof.

609.8 Structural Strength. Allowable stresses shall not be exceeded for materials used where a vertical or horizontal force of 250 pounds (1112 N) is applied at any point on the grab bar, fastener mounting device, or supporting structure.

610 Seats

610.1 General. Seats in accessible bathtubs and shower compartments shall comply with Section 610.

610.2 Bathtub Seats. The height of bathtub seats shall be 17 inches (430 mm) minimum and 19 inches (485 mm) maximum above the bathroom floor, measured to the top of the seat. Removable in-tub seats shall be 15 inches (380 mm) minimum and 16 inches (405 mm) maximum in depth. Removable in-tub seats shall be capable of secure placement. Permanent seats shall be 15 inches (380 mm) minimum in depth and shall extend from the back wall to or beyond the outer edge of the bathtub. Permanent seats shall be positioned at the head end of the bathtub.

610.3 Shower Compartment Seats. The height of shower compartment seats shall be 17 inches (430

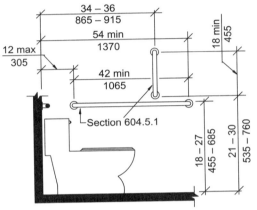

Note: For adult dimensions see Fig. 604.5.1
(a) Side Wall View

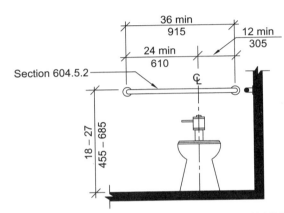

Note: For adult dimensions see Fig. 604.5.2
(b) Rear Wall View

**FIG. 609.4.2
POSITION OF CHILDREN'S GRAB BARS**

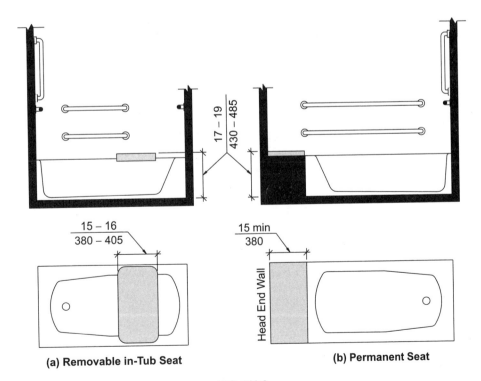

(a) Removable in-Tub Seat

(b) Permanent Seat

**FIG. 610.2
BATHTUB SEATS**

mm) minimum and 19 inches (485 mm) maximum above the bathroom floor, measured to the top of the seat. In transfer-type and alternate roll-in-type showers, the seat shall extend along the seat wall to a point within 3 inches (75 mm) of the compartment entry. In standard roll-in-type showers, the seat shall extend from the control wall to a point within 3 inches (75 mm) of the compartment entry. Seats shall comply with Section 610.3.1 or 610.3.2.

610.3.1 Rectangular Seats. The rear edge of a rectangular seat shall be $2^1/_2$ inches (64 mm) maximum and the front edge 15 inches (380 mm) minimum and 16 inches (405 mm) maximum from the seat wall. The side edge of the seat shall be $1^1/_2$ inches (38 mm) maximum from the back wall of a transfer-type shower and $1^1/_2$ inches (38 mm) maximum from the control wall of a roll-in-type shower.

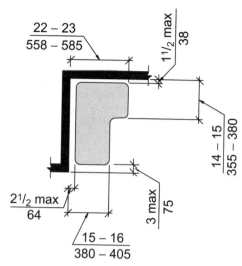

FIG. 610.3.2
L-SHAPED SHOWER COMPARTMENT SEAT

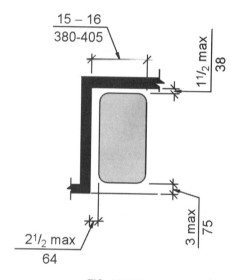

FIG. 610.3.1
RECTANGULAR SHOWER COMPARTMENT SEAT

610.3.2 L-Shaped Seats. The rear edge of an L-shaped seat shall be $2^1/_2$ inches (64 mm) maximum and the front edge 15 inches (380 mm) minimum and 16 inches (405 mm) maximum from the seat wall. The rear edge of the "L" portion of the seat shall be $1^1/_2$ inches (38 mm) maximum from the wall and the front edge shall be 14 inches (355 mm) minimum and 15 inches (380 mm) maximum from the wall. The end of the "L" shall be 22 inches (560 mm) minimum and 23 inches (585 mm) maximum from the main seat wall.

610.4 Structural Strength. Allowable stresses shall not be exceeded for materials used where a vertical or horizontal force of 250 pounds (1112 N) is applied at any point on the seat, fastener mounting device, or supporting structure.

611 Washing Machines and Clothes Dryers

611.1 General. Accessible washing machines and clothes dryers shall comply with Section 611.

611.2 Clear Floor Space. A clear floor space complying with Section 305, positioned for parallel approach, shall be provided. For top loading machines, the clear floor space shall be centered on the appliance. For front loading machines, the centerline of the clear floor space shall be offset 24 inches (610 mm) maximum from the centerline of the door opening.

611.3 Operable Parts. Operable parts, including doors, lint screens, detergent and bleach compartments, shall comply with Section 309.

611.4 Height. Top loading machines shall have the door to the laundry compartment 36 inches (915 mm) maximum above the floor. Front loading machines shall have the bottom of the opening to the laundry compartment 15 inches (380 mm) minimum and 36 inches (915 mm) maximum above the floor.

612 Saunas and Steam Rooms

612.1 General. Saunas and steam rooms shall comply with Section 612.

612.2 Bench. Where seating is provided in saunas and steam rooms, at least one bench shall comply with Section 903. Doors shall not swing into the clear floor space required by Section 903.2.

612.3 Turning space. A turning space complying with Section 304 shall be provided within saunas and steam rooms.

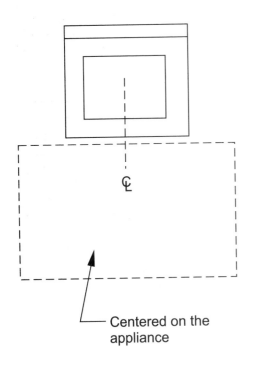

Centered on the appliance

(a) Top Loading

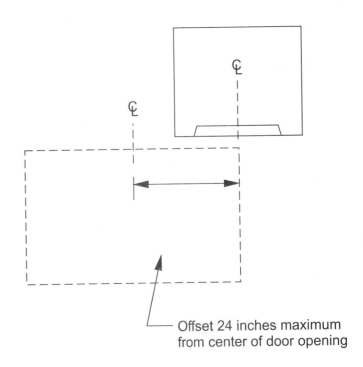

Offset 24 inches maximum from center of door opening

(b) Front Loading

FIG. 611.2
CLEAR FLOOR SPACE

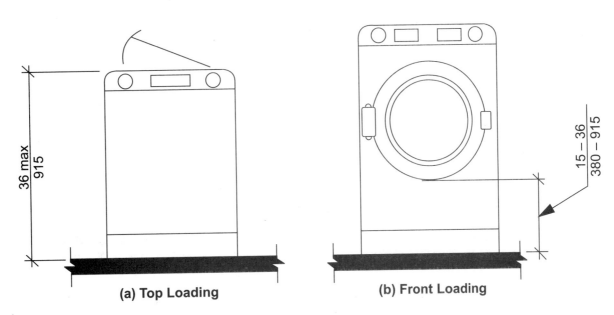

(a) Top Loading

(b) Front Loading

FIG. 611.4
HEIGHT OF LAUNDRY EQUIPMENT

Chapter 7. Communication Elements and Features

701 General

701.1 Scope. Communications elements and features required to be accessible by the scoping provisions adopted by the administrative authority shall comply with the applicable provisions of Chapter 7.

702 Alarms

702.1 General. Accessible audible and visible alarms and notification appliances shall be installed in accordance with NFPA 72 listed in Section 105.2.2, be powered by a commercial light and power source, be permanently connected to the wiring of the premises electric system, and be permanently installed.

703 Signs

703.1 General. Accessible signs shall comply with Section 703. Tactile signs shall contain both raised characters and braille. Where signs with both visual and raised characters are required, either one sign with both visual and raised characters, or two separate signs, one with visual, and one with raised characters, shall be provided.

703.1.1 Designations. Interior and exterior signs identifying permanent rooms and spaces shall comply with Sections 703.1, 703.2, and 703.3.

EXCEPTION: Exterior signs that are not located at the door to the space they serve shall not be required to comply with Section 703.3.

703.1.2 Directional and Informational Signs. Signs that provide direction to or information about interior spaces and facilities of the site shall comply with Section 703.2.

703.1.3 Pictograms. Where pictograms are provided as designations of permanent interior rooms and spaces, the pictograms shall comply with Section 703.5 and shall have text descriptors located directly below the pictogram field and complying with Sections 703.2 and 703.3.

EXCEPTION: Pictograms that provide information about a room or space, such as "No Smoking", occupant logos, and the International Symbol of Accessibility, are not required to have text descriptors.

703.2 Visual Characters.

703.2.1 General. Visual characters shall comply with the following:

1. Visual characters that also serve as raised characters shall comply with Section 703.3, or

2. Visual characters on VMS signage shall comply with Section 703.7, or

3. Visual characters not covered in items 1 and 2 shall comply with Section 703.2.

EXCEPTION: The visual and raised requirements of item 1 shall be permitted to be provided by two separate signs that provide corresponding information provided one sign complies with Section 703.2 and the second sign complies with Section 703.3.

703.2.2 Case. Characters shall be uppercase, lowercase, or a combination of both.

703.2.3 Style. Characters shall be conventional in form. Characters shall not be italic, oblique, script, highly decorative, or of other unusual forms.

703.2.4 Character Height. The uppercase letter "I" shall be used to determine the allowable height of all characters of a font. The uppercase letter "I" of the font shall have a minimum height complying with Table 703.2.4. Viewing distance shall be measured as the horizontal distance between the character and an

TABLE 703.2.4—VISUAL CHARACTER HEIGHT

Height above Floor to Baseline of Character	Horizontal Viewing Distance	Minimum Character Height
40 inches (1015 mm) to less than or equal to 70 inches (1780 mm)	Less than 6 feet (1830 mm)	$5/_8$ inch (16 mm)
	6 feet (1830 mm) and greater	$5/_8$ inch (16 mm), plus $1/_8$ inch (3.2 mm) per foot (305 mm) of viewing distance above 6 feet (1830 mm)
Greater than 70 inches (1780 mm) to less than or equal to 120 inches (3050 mm)	Less than 15 feet (4570 mm)	2 inches (51 mm)
	15 feet (4570 mm) and greater	2 inches (51 mm), plus $1/_8$ inch (3.2 mm) per foot (305 mm) of viewing distance above 15 feet (4570 mm)
Greater than 120 inches (3050 mm)	Less than 21 feet (6400 mm)	3 inches (75 mm)
	21 feet (6400 mm) and greater	3 inches (75 mm), plus $1/_8$ inch (3.2 mm) per foot (305 mm) of viewing distance above 21 feet (6400 mm)

obstruction preventing further approach towards the sign.

EXCEPTION: In assembly seating where the maximum viewing distance is 100 feet (30.5 m) or greater, the height of the uppercase "I" of fonts shall be permitted to be 1 inch (25 mm) for every 30 feet (9145 mm) of viewing distance, provided the character height is 8 inches (205 mm) minimum. Viewing distance shall be measured as the horizontal distance between the character and where someone is expected to view the sign.

703.2.5 Character Width. The uppercase letter "O" shall be used to determine the allowable width of all characters of a font. The width of the uppercase letter "O" of the font shall be 55 percent minimum and 110 percent maximum of the height of the uppercase "I" of the font.

703.2.6 Stroke Width. The uppercase letter "I" shall be used to determine the allowable stroke width of all characters of a font. The stroke width shall be 10 percent minimum and 30 percent maximum of the height of the uppercase "I" of the font.

703.2.7 Character Spacing. Spacing shall be measured between the two closest points of adjacent characters within a message, excluding word spaces. Spacing between individual characters shall be 10 percent minimum and 35 percent maximum of the character height.

703.2.8 Line Spacing. Spacing between the baselines of separate lines of characters within a message shall be 135 percent minimum and 170 percent maximum of the character height.

EXCEPTION: In assembly seating where the maximum viewing distance is 100 feet (30.5 m) or greater, the spacing between the baselines of separate lines of characters within a message shall be permitted to be 120 percent minimum and 170 percent maximum of the character height.

703.2.9 Height Above Floor. Visual characters shall be 40 inches (1015 mm) minimum above the floor of the viewing position, measured to the baseline of the character. Heights shall comply with Table 703.2.4, based on the size of the characters on the sign.

EXCEPTION: Visual characters indicating elevator car controls shall not be required to comply with Section 703.2.9.

703.2.10 Finish and Contrast. Characters and their background shall have a non-glare finish. Characters shall contrast with their background, with either light characters on a dark background, or dark characters on a light background.

703.3 Raised Characters.

703.3.1 General. Raised characters shall comply with Section 703.3, and shall be duplicated in braille complying with Section 703.4.

703.3.2 Depth. Raised characters shall be raised $^{1}/_{32}$ inch (0.8 mm) minimum above their background.

703.3.3 Case. Characters shall be uppercase.

703.3.4 Style. Characters shall be sans serif. Characters shall not be italic, oblique, script, highly decorative, or of other unusual forms.

703.3.5 Character Height. The uppercase letter "I" shall be used to determine the allowable height of all characters of a font. The height of the uppercase letter "I" of the font, measured vertically from the baseline of the character, shall be $^{5}/_{8}$ inch (16 mm) minimum, and 2 inches (51 mm) maximum.

EXCEPTION: Where separate raised and visual characters with the same information are provided, the height of the raised uppercase letter "I" shall be permitted to be $^{1}/_{2}$ inch (13 mm) minimum.

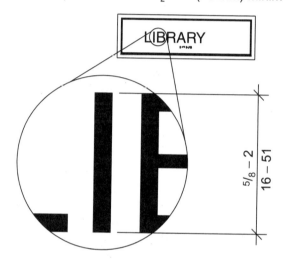

**FIG. 703.3.5
CHARACTER HEIGHT**

703.3.6 Character Width. The uppercase letter "O" shall be used to determine the allowable width of all characters of a font. The width of the uppercase letter "O" of the font shall be 55 percent minimum and 110 percent maximum of the height of the uppercase "I" of the font.

703.3.7 Stroke Width. Raised character stroke width shall comply with Section 703.3.7. The uppercase letter "I" of the font shall be used to determine the allowable stroke width of all characters of a font.

703.3.7.1 Maximum. The stroke width shall be 15 percent maximum of the height of the uppercase letter "I" measured at the top surface of the character, and 30 percent maximum of the height of the uppercase letter "I" measured at the base of the character.

703.3.7.2 Minimum. When characters are both visual and raised, the stroke width shall be 10 percent minimum of the height of the uppercase letter "I".

703.3.8 Character Spacing. Character spacing shall be measured between the two closest points of adjacent raised characters within a message, excluding word spaces. Spacing between individual raised characters shall be $^1/_8$ inch (3.2 mm) minimum measured at the top surface of the characters, $^1/_{16}$ inch (1.6 mm) minimum measured at the base of the characters, and four times the raised character stroke width maximum. Characters shall be separated from raised borders and decorative elements $^3/_8$ inch (9.5 mm) minimum.

703.3.9 Line Spacing. Spacing between the baselines of separate lines of raised characters within a message shall be 135 percent minimum and 170 percent maximum of the raised character height.

703.3.10 Height above Floor. Raised characters shall be 48 inches (1220 mm) minimum above the floor, measured to the baseline of the lowest raised character and 60 inches (1525 mm) maximum above

the floor, measured to the baseline of the highest raised character.

EXCEPTION: Raised characters for elevator car controls shall not be required to comply with Section 703.3.10.

703.3.11 Location. Where a sign containing raised characters and braille is provided at a door, the sign shall be alongside the door at the latch side. Where a sign containing raised characters and braille is provided at double doors with one active leaf, the sign shall be located on the inactive leaf. Where a sign containing raised characters and braille is provided at double doors with two active leaves, the sign shall be to the right of the right-hand door. Where there is no wall space on the latch side of a single door, or to the right side of double doors, signs shall be on the

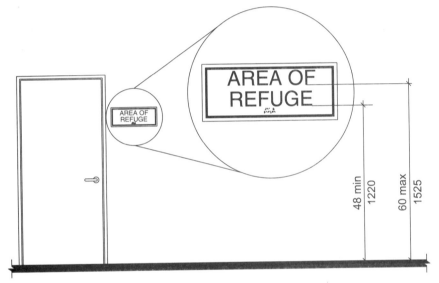

Note: For braille character mounting height see Section 703.4.5

FIG. 703.3.10
HEIGHT OF RAISED CHARACTERS ABOVE FLOOR

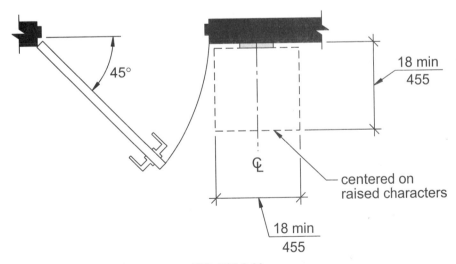

FIG. 703.3.11
LOCATION OF SIGNS AT DOORS

nearest adjacent wall. Signs containing raised characters and braille shall be located so that a clear floor area 18 inches (455 mm) minimum by 18 inches (455 mm) minimum, centered on the raised characters is provided beyond the arc of any door swing between the closed position and 45 degree open position.

EXCEPTION: Signs containing raised characters and braille shall be permitted on the push side of doors with closers and without hold-open devices.

703.3.12 Finish and Contrast. Characters and their background shall have a non-glare finish. Characters shall contrast with their background with either light characters on a dark background, or dark characters on a light background.

EXCEPTION: Where separate raised characters and visual characters with the same information are provided, raised characters are not required to have nonglare finish or to contrast with their background.

703.4 Braille

703.4.1 General. Braille shall be contracted (Grade 2) braille and shall comply with Section 703.4.

703.4.2 Uppercase Letters. The indication of an uppercase letter or letters shall only be used before the first word of sentences, proper nouns and names, individual letters of the alphabet, initials, or acronyms.

703.4.3 Dimensions. Braille dots shall have a domed or rounded shape and shall comply with Table 703.4.3.

703.4.4 Position. Braille shall be below the corresponding text. If text is multilined, braille shall be placed below entire text. Braille shall be separated $^3/_8$ inch (9.5 mm) minimum from any other raised characters and $^3/_8$ inch (9.5 mm) minimum from raised borders and decorative elements. Braille provided on elevator car controls shall be separated $^3/_{16}$ inch (4.8 mm) minimum either directly below or adjacent to the corresponding raised characters or symbols.

703.4.5 Mounting Height. Braille shall be 48 inches (1220 mm) minimum and 60 inches (1525 mm) maximum above the floor, measured to the baseline of the braille cells.

EXCEPTION: Elevator car controls shall not be required to comply with Section 703.4.5.

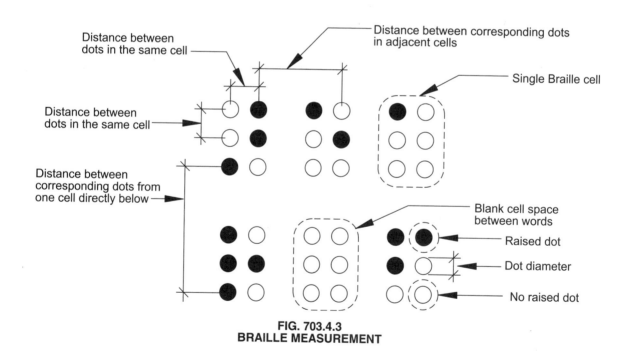

FIG. 703.4.3
BRAILLE MEASUREMENT

TABLE 703.4.3— BRAILLE DIMENSIONS

Measurement range	Minimum in inches Maximum in inches
Dot base diameter	0.059 (1.5 mm) to 0.063 (1.6 mm)
Distance between two dots in the same cell	0.090 (2.3 mm) to 0.100 (2.5 mm)
Distance between corresponding dots in adjacent cells[1]	0.241 (6.1 mm) to 0.300 (7.6 mm)
Dot height	0.025 (0.6 mm) to 0.037 (0.9 mm)
Distance between corresponding dots from one cell directly below[1]	0.395 (10.0 mm) to 0.400 (10.2 mm)

[1]Measured center to center

703.5 Pictograms.

703.5.1 General. Pictograms shall comply with Section 703.5.

703.5.2 Pictogram Field. Pictograms shall have a field 6 inches (150 mm) minimum in height. Characters or braille shall not be located in the pictogram field.

703.5.3 Finish and Contrast. Pictograms and their fields shall have a nonglare finish. Pictograms shall contrast with their fields, with either a light pictogram on a dark field or a dark pictogram on a light field.

703.6 Symbols of Accessibility.

703.6.1 General. Symbols of accessibility shall comply with Section 703.6.

7703.6.2 Finish and Contrast. Symbols of accessibility and their backgrounds shall have a non-glare finish. Symbols of accessibility shall contrast with their backgrounds, with either a light symbol on a dark background or a dark symbol on a light background.

703.6.3 Symbols.

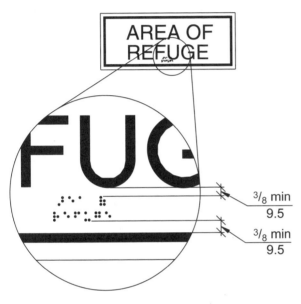

FIG. 703.4.4
POSITION OF BRAILLE

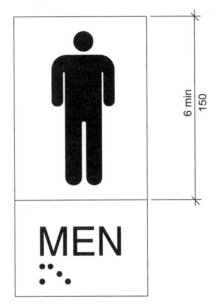

FIG. 703.5
PICTOGRAM FIELD

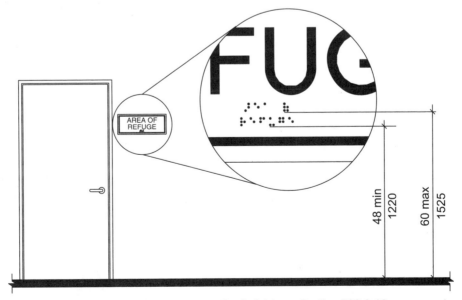

Note: For raised character mounting height see Section 703.3.10

FIG. 703.4.5
HEIGHT OF BRAILLE CHARACTERS ABOVE FLOOR

703.6.3.1 International Symbol of Accessibility. The International Symbol of Accessibility shall comply with Figure 703.6.3.1.

703.6.3.2 International Symbol of TTY. The International Symbol of TTY shall comply with Figure 703.6.3.2.

703.6.3.3 Assistive Listening Systems. Assistive listening systems shall be identified by the International Symbol of Access for Hearing Loss complying with Figure 703.6.3.3.

703.6.3.4 Volume-Controlled Telephones. Telephones with volume controls shall be identified by a pictogram of a telephone handset with radiating sound waves on a square field complying with Figure 703.6.3.4.

703.7 Variable Message Signs.

703.7.1 General. High resolution variable message sign (VMS) characters shall comply with Sections 703.2 and 703.7.12 through 703.7.14. Low resolution variable message sign (VMS) characters shall comply with Section 703.7.

EXCEPTION: Theatrical performance related VMS signs, including but not limited to, text and translation delivery systems, surtitles and subti-

tles, shall not be required to comply with Section 703.7.1.

703.7.2 Case. Low resolution VMS characters shall be uppercase.

703.7.3 Style. Low resolution VMS characters shall be conventional in form, shall be san serif, and shall not be italic, oblique, script, highly decorative, or of other unusual forms.

703.7.4 Character Height. The uppercase letter "I" shall be used to determine the allowable height of all low resolution VMS characters of a font. Viewing distance shall be measured as the horizontal distance between the character and an obstruction preventing further approach towards the sign. The uppercase letter "I" of the font shall have a minimum height complying with Table 703.7.4.

EXCEPTION: In assembly seating where the maximum viewing distance is 100 feet (30.5 m) or greater, the height of the uppercase "I" of low resolution VMS fonts shall be permitted to be 1 inch (25 mm) for every 30 feet (9145 mm) of viewing distance, provided the character height is 8 inches (205 mm) minimum. Viewing distance shall be measured as the horizontal distance between the character and where someone is expected to view the sign.

FIG. 703.6.3.1
INTERNATIONAL SYMBOL OF ACCESSIBILITY

FIG. 703.6.3.3
INTERNATIONAL SYMBOL OF
ACCESS FOR HEARING LOSS

FIG. 703.6.3.2
INTERNATIONAL TTY SYMBOL

FIG. 703.6.3.4
VOLUME-CONTROLLED TELEPHONE

703.7.5 Character Width. The uppercase letter "O" shall be used to determine the allowable width of all low resolution VMS characters of a font. Low resolution VMS characters shall comply with the pixel count for character width in Table 703.7.5.

703.7.6 Stroke Width. The uppercase letter "I" shall be used to determine the allowable stroke width of all low resolution VMS characters of a font. Low resolution VMS characters shall comply with the pixel count for stroke width in Table 703.7.5.

703.7.7 Character Spacing. Spacing shall be measured between the two closest points of adjacent low resolution VMS characters within a message, excluding word spaces. Low resolution VMS character spacing shall comply with the pixel count for character spacing in Table 703.7.5.

703.7.8 Line Spacing. Low resolution VMS characters shall comply with Section 703.2.8.

703.7.9 Height Above Floor. Low resolution VMS characters shall be 40 inches (1015 mm) minimum above the floor of the viewing position, measured to the baseline of the character. Heights of low resolution variable message sign characters shall comply with Table 703.7.4, based on the size of the characters on the sign.

703.7.10 Finish. The background of Low resolution VMS characters shall have a non-glare finish.

703.7.11 Contrast. Low resolution VMS characters shall be light characters on a dark background.

703.7.12 Protective Covering. Where a protective layer is placed over VMS characters through which the VMS characters must be viewed, the protective covering shall have a non-glare finish.

703.7.13 Brightness. The brightness of variable message signs in exterior locations shall automatically adjust in response to changes in ambient light levels.

703.7.14 Rate of Change. Where a VMS message can be displayed in its entirety on a single screen, it shall be displayed on a single screen and shall remain motionless on the screen for a minimum 3 seconds or one second minimum for every 7 characters of the message including spaces whichever is longer.

TABLE 703.7.4—LOW RESOLUTION VMS CHARACTER HEIGHT

Height above Floor to Baseline of Character	Horizontal Viewing Distance	Minimum Character Height
40 inches (1015 mm) to less than or equal to 70 inches (1780 mm)	Less than 10 feet (3050 mm)	2 inches (51 mm)
	10 feet (3050 mm) and greater	2 inches (51 mm), plus 1/5 inch (5.1 mm) per foot (305 mm) of viewing distance above 10 feet (3050 mm)
Greater than 70 inches (1780 mm) to less than or equal to 120 inches (3050 mm)	Less than 15 feet (4570 mm)	3 inches (75 mm)
	15 feet (4570 mm) and greater	3 inches (75 mm), plus 1/5 inch (5.1 mm) per foot (305 mm) of viewing distance above 15 feet (4570 mm)
Greater than 120 inches (3050 mm)	Less than 20 feet (6095 mm)	4 inches (100 mm)
	20 feet (6095 mm) and greater	4 inches (100 mm), plus 1/5 inch (5.1 mm) per foot (305 mm) of viewing distance above 20 feet (6095 mm)

TABLE 703.7.5 PIXEL COUNT FOR LOW RESOLUTION VMS SIGNAGE[1]

Character Height	Character Width Range	Stroke Width Range	Character Spacing Range
7	5-6	1	2
8	6-7	1-2	2-3
9	6-8	1-2	2-3
10	7-9	2	2-4
11	8-10	2	2-4
12	8-11	2	3-4
13	9-12	2-3	3-5
14	10-13	2-3	3-5
15	11-14	2-3	3-5

(1) Measured in pixels.

Example 1

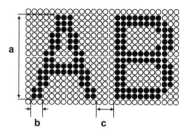

Example 2

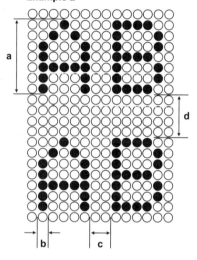

	Property	Example 1	Example 2
a	Character Height	14 Pixels	7 Pixels
b	Stroke Width	2 Pixels	1 Pixel
c	Character Spacing	3 Pixels	2 Pixels
d	Line Spacing		4 Pixels

FIG. 703.7.5
LOW RESOLUTION VMS SIGNAGE CHARACTERS

703.8 Remote Infrared Audible Sign (RIAS) Systems.

703.8.1 General. Remote Infrared Audible Sign Systems shall comply with Section 703.8.

703.8.2 Transmitters. Where provided, Remote Infrared Audible Sign Transmitters shall be designed to communicate with receivers complying with Section 703.8.3.

703.8.3 Infrared Audible Sign Receivers.

703.8.3.1 Frequency. Basic speech messages shall be frequency modulated at 25 kHz, with a +/- 2.5 kHz deviation, and shall have an infrared wavelength from 850 to 950 nanometer (nm).

703.8.3.2 Optical Power Density. Receiver shall produce a 12 decibel (dB) signal-plus-noise-to-

noise ratio with a 1 kHz modulation tone at +/- 2.5 kHz deviation of the 25 kHz subcarrier at an optical power density of 26 picowatts per square millimeter measured at the receiver photosensor aperture.

703.8.3.3 Audio Output. The audio output from an internal speaker shall be at 75 dBA minimum at 18 inches (455 mm) with a maximum distortion of 10 percent.

703.8.3.4 Reception Range. The receiver shall be designed for a high dynamic range and capable of operating in full-sun background illumination.

703.8.3.5 Multiple Signals. A receiver provided for the capture of the stronger of two signals in the receiver field of view shall provide a received power ratio on the order of 20 dB for negligible interference.

703.9 Pedestrian Signals. Accessible pedestrian signals shall comply with Section 4E.06-Accessible Pedestrian Signals, and Section 4E.09-Accessible Pedestrian Signal Detectors, of the Manual on Uniform Traffic Control Devices listed in Section 105.2.1.

EXCEPTION: Pedestrian signals are not required to comply with the requirement for choosing audible tones.

704 Telephones.

704.1 General. Accessible public telephones shall comply with Section 704.

704.2 Wheelchair Accessible Telephones. Wheelchair accessible public telephones shall comply with Section 704.2.

EXCEPTION: Drive up only public telephones are not required to comply with Section 704.2.

704.2.1 Clear Floor Space. A clear floor space complying with Section 305 shall be provided. The clear floor space shall not be obstructed by bases, enclosures, or seats.

704.2.1.1 Parallel Approach. Where a parallel approach is provided, the distance from the edge of the telephone enclosure to the face of the telephone shall be 10 inches (255 mm) maximum.

704.2.1.2 Forward Approach. Where a forward approach is provided, the distance from the front edge of a counter within the enclosure to the face of the telephone shall be 20 inches (510 mm) maximum.

704.2.2 Operable Parts. Operable parts shall comply with Section 309. Telephones shall have push button controls where service for such equipment is available.

704.2.3 Telephone Directories. Where provided, telephone directories shall comply with Section 309.

704.2.4 Cord Length. The telephone handset cord shall be 29 inches (735 mm) minimum in length.

704.2.5 Hearing-Aid Compatibility. Telephones shall be hearing aid compatible.

704.3 Volume-control Telephones. Public telephones required to have volume controls shall be equipped with a receiver volume control that provides a gain adjustable up to 20 dB minimum. Incremental volume controls shall provide at least one intermediate step of gain of 12 dB minimum. An automatic reset shall be provided.

704.4 TTY. TTYs required at a public pay telephone shall be permanently affixed within, or adjacent to, the telephone enclosure. Where an acoustic coupler is used, the telephone cord shall be of sufficient length to allow connection of the TTY and the telephone receiver.

704.5 Height. When in use, the touch surface of TTY keypads shall be 34 inches (865 mm) minimum above the floor.

> **EXCEPTION:** Where seats are provided, TTYs shall not be required to comply with Section 704.5.

704.6 TTY Shelf. Where public pay telephones designed to accommodate a portable TTY are provided, they shall be equipped with a shelf and an electrical outlet within or adjacent to the telephone enclosure. The telephone handset shall be capable of being placed flush on the surface of the shelf. The shelf shall be capable of accommodating a TTY and shall have a vertical clearance 6 inches (150 mm) minimum in height above the area where the TTY is placed.

704.7 Protruding Objects. Telephones, enclosures, and related equipment shall comply with Section 307.

705 Detectable Warnings

705.1 General. Detectable warning surfaces shall comply with Section 705.

705.2 Standardization. Detectable warning surfaces shall be standard within a building, facility, site, or complex of buildings.

> **EXCEPTION:** In facilities that have both interior and exterior locations, detectable warnings in exterior locations shall not be required to comply with Section 705.4.

705.3 Contrast. Detectable warning surfaces shall contrast visually with adjacent surfaces, either light-on-dark or dark-on-light.

705.4 Interior Locations. Detectable warning surfaces in interior locations shall differ from adjoining walking surfaces in resiliency or sound-on-cane contact.

705.5 Truncated Domes. Detectable warning surfaces shall have truncated domes complying with Section 705.5.

705.5.1 Size. Truncated domes shall have a base diameter of 0.9 inch (23 mm) minimum and 1.4 inch (36 mm) maximum, and a top diameter of 50 percent minimum and 65 percent maximum of the base diameter.

705.5.2 Height. Truncated domes shall have a height of 0.2 inch (5.1 mm).

705.5.3 Spacing. Truncated domes shall have a center-to-center spacing of 1.6 inches (41 mm) minimum and 2.4 inches (61 mm) maximum, and a base-to-base spacing of 0.65 inch (16.5 mm) minimum, measured between the most adjacent domes on the grid.

705.5.4 Alignment. Truncated domes shall be aligned in a square grid pattern.

705.6 Transportation Platform Edges. Detectable warning surfaces at transportation platform boarding edges shall extend the full length of the public use areas of the platform. The detectable warning surface shall extend 24 inches (610 mm) from the boarding edge of the platform.

706 Assistive Listening Systems

706.1 General. Accessible assistive listening systems in assembly areas shall comply with Section 706.

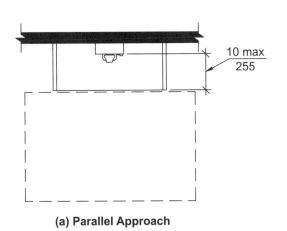

(a) Parallel Approach

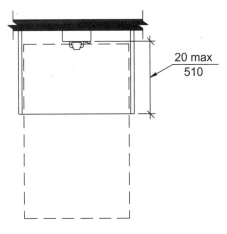

(b) Forward Approach

FIG. 704.2.1
CLEAR FLOOR SPACE FOR TELEPHONES

706.2 Receiver Jacks. Receivers required for use with an assistive listening system shall include a $\frac{1}{8}$ inch (3.2 mm) standard mono jack.

706.3 Receiver Hearing-aid Compatibility. Receivers required to be hearing aid compatible shall interface with telecoils in hearing aids through the provision of neck loops.

706.4 Sound Pressure Level. Assistive listening systems shall be capable of providing a sound pressure level of 110 dB minimum and 118 dB maximum, with a dynamic range on the volume control of 50 dB.

706.5 Signal-to-noise Ratio. The signal-to-noise ratio for internally generated noise in assistive listening systems shall be 18 dB minimum.

706.6 Peak Clipping Level. Peak clipping shall not exceed 18 dB of clipping relative to the peaks of speech.

707 Automatic Teller Machines (ATMs) and Fare Machines

707.1 General. Accessible automatic teller machines and fare machines shall comply with Section 707.

707.2 Clear Floor Space. A clear floor space complying with Section 305 shall be provided in front of the machine.

> **EXCEPTION:** Clearfloor space is not required at drive up only automatic teller machines and fare machines.

707.3 Operable Parts. Operable parts shall comply with Section 309. Unless a clear or correct key is provided, each operable part shall be able to be differentiated by sound or touch, without activation.

> **EXCEPTION:** Drive up only automatic teller machines and fare machines shall not be required to comply with Section 309.2 or 309.3.

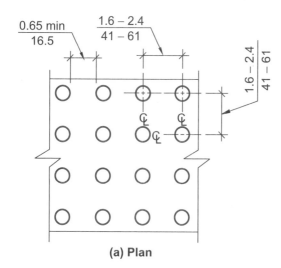 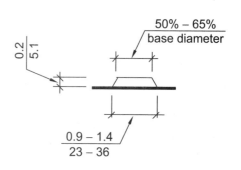

(a) Plan (b) Elevation (Enlarged)

FIG. 705.5
TRUNCATED DOME SIZE AND SPACING

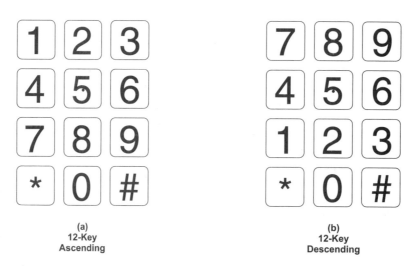

(a)
12-Key
Ascending

(b)
12-Key
Descending

FIG. 707.5
NUMERIC KEY LAYOUT

707.4 Privacy. Automatic teller machines shall provide the opportunity for the same degree of privacy of input and output available to all individuals.

707.5 Numeric Keys. Numeric keys shall be arranged in a 12-key ascending or descending telephone keypad layout. The number Five key shall have a single raised dot.

707.6 Function Keys. Function keys shall comply with Section 707.6.

707.6.1 Raised Symbols. Function key surfaces shall have raised symbols as shown in Table 707.6.1.

TABLE 707.6.1—RAISED SYMBOLS

Key Function	Description of Raised Symbol	Raised Symbol
Enter or Proceed:	CIRCLE	◯
Clear or Correct:	LEFT ARROW	←
Cancel:	"X"	X
Add Value:	PLUS SIGN	+
Decreased Value:	MINUS SIGN	-

707.6.2 Contrast. Function keys shall contrast visually from background surfaces. Characters and symbols on key surfaces shall contrast visually from key surfaces. Visual contrast shall be either light-on-dark or dark-on-light.

EXCEPTION: Raised symbols required by Section 707.6.1 shall not be required to comply with Section 707.6.2.

707.7 Display Screen. The display screen shall comply with Section 707.7.

707.7.1 Visibility. The display screen shall be visible from a point located 40 inches (1015 mm) above the center of the clear floor space in front of the machine.

EXCEPTION: Drive up only automatic teller machines and fare machines shall not be required to comply with Section 707.7.1.

707.7.2 Characters. Characters displayed on the screen shall be in a sans serif font. The uppercase letter "I" shall be used to determine the allowable height of all characters of the font. The uppercase letter "I" of the font shall be $^3/_{16}$ inch (4.8 mm) minimum in height. Characters shall contrast with their background with either light characters on a dark background, or dark characters on a light background.

707.8 Speech Output. Machines shall be speech enabled. Operating instructions and orientation, visible transaction prompts, user input verification, error messages, and all displayed information for full use shall be accessible to and independently usable by individuals

with vision impairments. Speech shall be delivered through a mechanism that is readily available to all users including, but not limited to, an industry standard connector or a telephone handset. Speech shall be recorded or digitized human, or synthesized.

EXCEPTIONS:

1. Audible tones shall be permitted in lieu of speech for visible output that is not displayed for security purposes, including but not limited to, asterisks representing personal identification numbers.

2. Advertisements and other similar information shall not be required to be audible unless they convey information that can be used in the transaction being conducted.

3. Where speech synthesis cannot be supported, dynamic alphabetic output shall not be required to be audible.

707.8.1 User Control. Speech shall be capable of being repeated and interrupted by the user. There shall be a volume control for the speech function.

EXCEPTION: Speech output for any single function shall be permitted to be automatically interrupted when a transaction is selected.

707.8.2 Receipts. Where receipts are provided, speech output devices shall provide audible balance inquiry information, error messages, and all other information on the printed receipt necessary to complete or verify the transaction.

EXCEPTIONS:

1. Machine location, date and time of transaction, customer account number, and the machine identifier shall not be required to be audible.

2. Information on printed receipts that duplicates audible information available on-screen shall not be required to be presented in the form of an audible receipt.

3. Printed copies of bank statements and checks shall not be required to be audible.

707.9 Input Controls. At least one tactually discernible input control shall be provided for each function. Where provided, key surfaces not on active areas of display screens shall be raised above surrounding surfaces. Where membrane keys are the only method of input, each shall be tactually discernable from surrounding surfaces and adjacent keys.

707.10 Braille Instructions. Braille instructions for initiating the speech mode shall be provided. Braille shall comply with Section 703.4.

708 Two-way Communication Systems

708.1 General. Accessible two-way communication systems shall comply with Section 708.

708.2 Audible and Visual Indicators. The system shall provide both visual and audible signals.

708.3 Handsets. Handset cords, if provided, shall be 29 inches (735 mm) minimum in length.

708.4 Telephone entry systems. Telephone entry systems shall comply with ANSI/DASMA 303 listed in Section 105.2.7.

Chapter 8. Special Rooms and Spaces

801 General

801.1 Scope. Special rooms and spaces required to be accessible by the scoping provisions adopted by the administrative authority shall comply with the applicable provisions of Chapter 8.

802 Assembly Areas

802.1 General. Wheelchair spaces and wheel chair space locations in assembly areas with spectator seating shall comply with Section 802. Team and player seating shall comply with Sections 802.2 through 802.6.

802.2 Floor Surfaces. The floor surface of wheelchair space locations shall have a slope not steeper than 1:48 and shall comply with Section 302.

802.3 Width. A single wheelchair space shall be 36 inches (915 mm) minimum in width. Where two adjacent wheelchair spaces are provided, each wheelchair space shall be 33 inches (840 mm) minimum in width.

802.4 Depth. Where a wheelchair space can be entered from the front or rear, the wheelchair space shall be 48 inches (1220 mm) minimum in depth. Where a wheelchair space can only be entered from the side, the wheelchair space shall be 60 inches (1525 mm) minimum in depth.

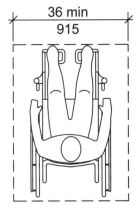

(a) Single Space

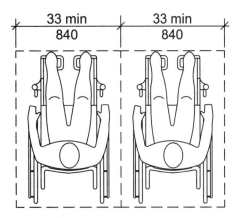

(b) Multiple Adjacent Spaces

FIG. 802.3
WIDTH OF A WHEELCHAIR SPACE IN ASSEMBLY AREAS

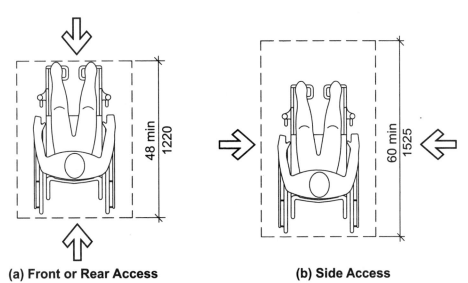

(a) Front or Rear Access   **(b) Side Access**

FIG. 802.4
DEPTH OF A WHEELCHAIR SPACE IN ASSEMBLY AREAS

802.5 Approach. The wheelchair space shall adjoin an accessible route. The accessible route shall not overlap the wheelchair space.

802.5.1 Overlap. A wheelchair space shall not overlap the required width of an aisle.

802.6 Integration of Wheelchair Space Locations. Wheelchair space locations shall be an integral part of any seating area.

802.7 Companion Seat. A companion seat, complying with Section 802.7, shall be provided beside each wheelchair space.

802.7.1 Companion Seat Type. The companion seat shall be equivalent in size, quality, comfort and amenities to the seats in the immediate area to the wheelchair space location. Companion seats shall be permitted to be moveable.

802.7.2 Companion Seat Alignment. In row seating, the companion seat shall be located to provide shoulder alignment with the wheelchair space occupant. The shoulder of the wheelchair space occupant shall be measured either 36 inches (915 mm) from the front or 12 inches (305 mm) from the rear of the wheelchair space. The floor surface for the companion seat shall be at the same elevation as the wheelchair space floor surface.

802.8 Designated Aisle Seats. Designated aisle seats shall comply with Section 802.8.

802.8.1 Armrests. Where armrests are provided on seating in the immediate area of designated aisle seats, folding or retractable armrests shall be provided on the aisle side of the designated aisle seat.

802.8.2 Identification. Each designated aisle seat shall be identified by the International Symbol of Accessibility.

802.9 Lines of Sight. Where spectators are expected to remain seated for purposes of viewing events, spectators in wheelchair space locations shall be provided with a line of sight in accordance with Section 802.9.1. Where spectators in front of the wheelchair space locations will be expected to stand at their seats for purposes of viewing events, spectators in wheelchair space locations shall be provided with a line of sight in accordance with Section 802.9.2.

802.9.1 Line of Sight over Seated Spectators. Where spectators are expected to remain seated during events, spectators seated in a wheelchair space shall be provided with lines of sight to the performance area or playing field comparable to that provided to seated spectators in closest proximity to the wheelchair space location. Where seating provides lines of sight over heads, spectators in wheelchair space locations shall be afforded lines of sight complying with Section 802.9.1.1. Where wheelchair space locations provide lines of sight over the shoulder and between heads, spectators in wheelchair space locations shall be afforded lines of sight complying with Section 802.9.1.2.

802.9.1.1 Lines of Sight over Heads. Spectators seated in a wheelchair space shall be afforded lines of sight over the heads of seated individuals in the first row in front of the wheelchair space location.

802.9.1.2 Lines of Sight between Heads. Spectators seated in a wheelchair space shall be afforded lines of sight over the shoulders and between the heads of seated individuals in the first row in front of the wheelchair space location.

802.9.2 Line of Sight over Standing Spectators. Wheelchair spaces required to provide a line of sight over standing spectators shall comply with Section 802.9.2.

802.9.2.1 Distance from Adjacent Seating. The front of the wheelchair space in a wheelchair space location shall be 12 inches (305 mm) maximum from the back of the chair or bench in front.

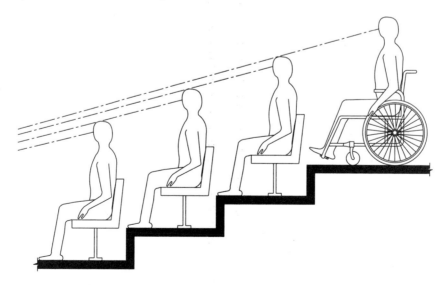

FIG. 802.9.1.1
LINES OF SIGHT OVER THE HEADS OF SEATED SPECTATORS

802.9.2.2 Height. The height of the floor surface at the wheelchair space location shall comply with Table 802.9.2.2. Interpolations shall be permitted for riser heights that are not listed in the table.

802.10 Wheelchair Space Dispersion. The minimum number of wheelchair space locations shall be in accordance with Table 802.10. Wheelchair space locations shall be dispersed in accordance with Sections 802.10.1, 802.10.2 and 802.10.3. In addition, wheelchair space locations shall be dispersed in accordance with Section 802.10.4 in spaces utilized primarily for viewing motion picture projection. Once the required number of wheelchair space locations has been met, further dispersion is not required.

802.10.1 Horizontal Dispersion. Wheelchair space locations shall be dispersed horizontally to provide viewing options. Two wheelchair spaces shall be permitted to be located side-by-side.

EXCEPTION: Horizontal dispersion shall not be required in assembly areas with 300 or fewer seats if the wheelchair space locations are located within the 2nd and 3rd quartile of the row length.

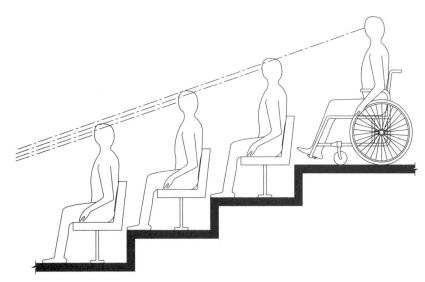

FIG. 802.9.1.2
LINES OF SIGHT BETWEEN THE HEADS OF SEATED SPECTATORS

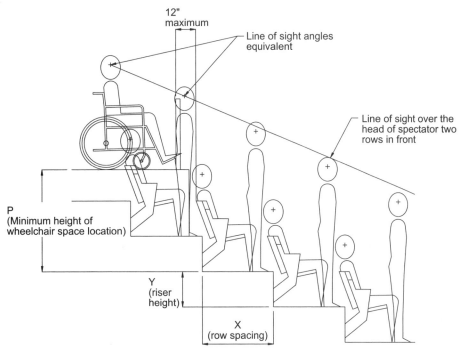

FIG. 802.9.2
LINE OF SIGHT OVER STANDING SPECTATORS

Intermediate aisles shall be included in determining the total row length. If the row length in the 2nd and 3rd quartile of the row is insufficient to accommodate the required number of companion seats and wheelchair spaces, the additional companion seats and wheelchair spaces shall be permitted to extend into in the 1st and 4th quartile of the row.

802.10.2 Dispersion for Variety of Distances from the Event. Wheelchair space locations shall be dispersed at a variety of distances from the event to provide viewing options.

EXCEPTIONS:

1. In bleachers, wheelchair space locations provided only in rows at points of entry to bleacher seating shall be permitted.

2. Assembly areas utilized for viewing motion picture projections with 300 seats or less shall not be required to comply with Section 802.10.2.

3. Assembly areas with 300 seats or less other than those utilized for viewing motion picture projections shall not be required to comply with Section 802.10.2 where all wheelchair space locations are within the front 50 percent of the total rows.

802.10.3 Dispersion by Type. Where assembly seating has multiple distinct seating areas with amenities that differ from other distinct seating areas, wheelchair space locations shall be provided within each distinct seating area.

802.10.4 Spaces Utilized Primarily for Viewing Motion Picture Projections. In spaces utilized primarily for viewing motion picture projections, wheelchair space locations shall comply with Section 802.10.4.

802.10.4.1 Spaces with Seating on Risers. Where tiered seating is provided, wheelchair space locations shall be integrated into the tiered seating area.

TABLE 802.9.2.2
REQUIRED WHEELCHAIR SPACE LOCATION ELEVATION OVER STANDING SPECTATORS

Riser height	Minimum height of the wheelchair space location based on row spacing[1]		
	Rows less than 33 inches (840 mm)[2]	Rows 33 inches (840 mm) to 44 inches (1120 mm)[2]	Rows over 44 inches (1120 mm)[2]
0 inch (0 mm)	16 inches (405 mm)	16 inches (405 mm)	16 inches (405 mm)
4 inches (100 mm)	22 inches (560 mm)	21 inches (535 mm)	21 inches (535 mm)
8 inches (205 mm)	31 inches (785 mm)	30 inches (760 mm)	28 inches (710 mm)
12 inches (305 mm)	40 inches (1015 mm)	37 inches (940 mm)	35 inches (890 mm)
16 inches (405 mm)	49 inches (1245 mm)	45 inches (1145 mm)	42 inches (1065 mm)
20 inches (510 mm)[3]	58 inches (1475 mm)	53 inches (1345 mm)	49 inches (1245 mm)
24 inches (610 mm)	N/A	61 inches (1550 mm)	56 inches (1420 mm)
28 inches (710 mm)[4]	N/A	69 inches (1750 mm)	63 inches (1600 mm)
32 inches (815 mm)	N/A	N/A	70 inches (1780 mm)
36 inches (915 mm) and higher	N/A	N/A	77 inches (1955 mm)

Footnotes to Table 802.9.2.2

[1]The height of the wheelchair space location is the vertical distance from the tread of the row of seats directly in front of the wheelchair space location to the tread of the wheelchair space location.

[2]The row spacing is the back-to-back horizontal distance between the rows of seats in front of the wheelchair space location.

[3]Seating treads less than 33 inches (840 mm) in depth are not permitted with risers greater than 18 inches (455 mm) in height.

[4]Seating treads less than 44 inches (1120 mm) in depth are not permitted with risers greater than 27 inches (685 mm) in height.

NOTE: Table 802.9.2.2 is based on providing a spectator in a wheelchair a line of sight over the head of a spectator two rows in front of the wheelchair space location using average anthropometrical data. The table is based on the following calculation: $[(2X+34)(Y-2.25)/X]+(20.2-Y)$ where Y is the riser height of the rows in front of the wheelchair space location and X is the tread depth of the rows in front of the wheelchair space location. The calculation is based on the front of the wheelchair space location being located 12 inches (305 mm) from the back of the seating tread directly in front and the eye of the standing spectator being set back 8 inches (205 mm) from the riser.

TABLE 802.10
WHEELCHAIR SPACE LOCATION DISPERSION

Total seating in Assembly Areas	Minimum required number of wheelchair space locations
Up to 150	1
151 to 500	2
501 to 1000	3
1001 to 5,000	3, plus 1 additional space for each 1,000 seats or portions thereof above 1,000
5,001 and over	7, plus 1 additional space for each 2,000 seats or portions thereof above 5,000

802.10.4.2 Distance from the Screen. Wheelchair space locations shall be located within the rear 60 percent of the seats provided.

803 Dressing, Fitting, and Locker Rooms

803.1 General. Accessible dressing, fitting, and locker rooms shall comply with Section 803.

803.2 Turning Space. A turning space complying with Section 304 shall be provided within the room.

803.3 Door Swing. Doors shall not swing into the room unless a clear floor space complying with Section 305.3 is provided within the room, beyond the arc of the door swing.

803.4 Benches. A bench complying with Section 903 shall be provided within the room.

803.5 Coat Hooks and Shelves. Accessible coat hooks provided within the room shall accommodate a forward reach or side reach complying with Section 308. Where provided, a shelf shall be 40 inches (1015 mm) minimum and 48 inches (1220 mm) maximum above the floor.

804 Kitchens and Kitchenettes

804.1 General. Accessible kitchens and kitchenettes shall comply with Section 804.

804.2 Clearance. Where a pass-through kitchen is provided, clearances shall comply with Section 804.2.1.

Where a U-shaped kitchen is provided, clearances shall comply with Section 804.2.2.

EXCEPTION: Spaces that do not provide a cooktop or conventional range shall not be required to comply with Section 804.2 provided there is a 40-inch (1015 mm) minimum clearance between all opposing base cabinets, counter tops, appliances, or walls within work areas.

804.2.1 Pass-through Kitchens. In pass-through kitchens where counters, appliances or cabinets are on two opposing sides, or where counters, appliances or cabinets are opposite a parallel wall, clearance between all opposing base cabinets, counter tops, appliances, or walls within kitchen work areas shall be 40 inches (1015 mm) minimum. Pass-through kitchens shall have two entries.

804.2.2 U-Shaped Kitchens. In kitchens enclosed on three contiguous sides, clearance between all opposing base cabinets, countertops, appliances, or walls within kitchen work areas shall be 60 inches (1525 mm) minimum.

804.3 Work Surface. At least one work surface shall be provided in accordance with Section 902.

EXCEPTION: Spaces that do not provide a cooktop or conventional range shall not be required to provide an accessible work surface.

804.4 Sinks. The sink shall comply with Section 606.

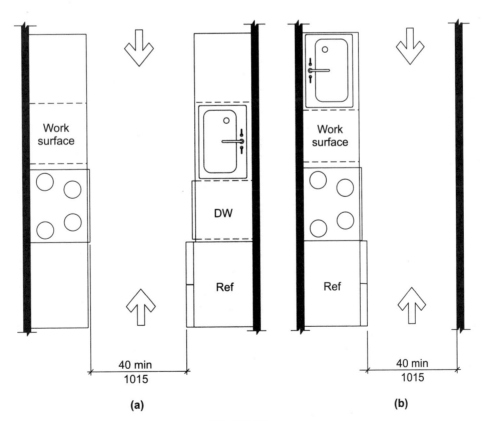

FIG. 804.2.1
PASS-THROUGH KITCHEN CLEARANCE

804.5 Appliances. Where provided, kitchen appliances shall comply with Section 804.5.

804.5.1 Clear Floor Space. A clear floor space complying with Section 305 shall be provided at each kitchen appliance.

804.5.2 Operable Parts. All appliance controls shall comply with Section 309.

EXCEPTIONS:

1. Appliance doors and door latching devices shall not be required to comply with Section 309.4.

2. Bottom-hinged appliance doors, when in the open position, shall not be required to comply with Section 309.3.

804.5.3 Dishwasher. A clear floor space positioned adjacent to the dishwasher door, shall be provided. The dishwasher door in the open position shall not obstruct the clear floor space for the dishwasher or an adjacent sink.

804.5.4 Cooktop. Cooktops shall comply with Section 804.5.4.

804.5.4.1 Approach. A clear floor space, positioned for a parallel or forward approach to the cooktop, shall be provided.

804.5.4.2 Forward approach. Where the clear floor space is positioned for a forward approach, knee and toe clearance complying with Section 306 shall be provided. The underside of the cooktop shall be insulated or otherwise configured to prevent burns, abrasions, or electrical shock.

804.5.4.3 Parallel approach. Where the clear floor space is positioned for a parallel approach, the clear floor space shall be centered on the appliance.

804.5.4.4 Controls. The location of controls shall not require reaching across burners.

804.5.5 Oven. Ovens shall comply with Section 804.5.5.

804.5.5.1 Clear floor space. A clear floor space shall be provided. The oven door in the open position shall not obstruct the clear floor space for the oven.

804.5.5.2 Side-Hinged Door Ovens. Side-hinged door ovens shall have a work surface complying with Section 804.3 positioned adjacent to the latch side of the oven door.

804.5.5.3 Bottom-Hinged Door Ovens. Bottom-hinged door ovens shall have a work surface complying with Section 804.3 positioned adjacent to one side of the door.

804.5.5.4 Controls. The location of controls shall not require reaching across burners.

804.5.6 Refrigerator/Freezer. Combination refrigerators and freezers shall have at least 50 percent of the freezer compartment shelves, including the bottom of the freezer, 54 inches (1370 mm) maximum above the floor when the shelves are installed at the maximum heights possible in the compartment. A clear floor space, positioned for a parallel approach to the refrigerator/freezer, shall be provided. The centerline of the

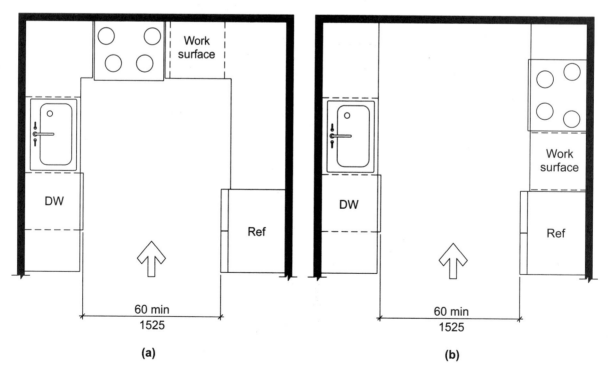

(a)

(b)

FIG. 804.2.2
U-SHAPED KITCHEN CLEARANCE

clear floor space shall be offset 24 inches (610 mm) maximum from the centerline of the appliance.

805 Transportation Facilities

805.1 General. Transportation facilities shall comply with Section 805.

805.2 Bus Boarding and Alighting Areas. Bus boarding and alighting areas shall comply with Section 805.2.

805.2.1 Surface. Bus stop boarding and alighting areas shall have a firm, stable surface.

805.2.2 Dimensions. Bus stop boarding and alighting areas shall have a 96-inch (2440 mm) minimum clear length, measured perpendicular to the curb or vehicle roadway edge, and a 60-inch (1525 mm) minimum clear width, measured parallel to the vehicle roadway.

805.2.3 Slope. The slope of the bus stop boarding and alighting area parallel to the vehicle roadway shall be the same as the roadway, to the maximum extent practicable. The slope of the bus stop boarding and alighting area perpendicular to the vehicle roadway shall be 1:48 maximum.

805.2.4 Connection. Bus stop boarding and alighting areas shall be connected to streets, sidewalks, or pedestrian paths by an accessible route complying with Section 402.

805.3 Bus Shelters. Bus shelters shall provide a minimum clear floor space complying with Section 305 entirely within the shelter. Bus shelters shall be connected by an accessible route complying with Section 402 to a boarding and alighting area complying with Section 805.2.

805.4 Bus Signs. Bus route identification signs shall have visual characters complying with Sections 703.2.2, 703.2.3, and 703.2.5 through 703.2.8. In addition, bus route identification numbers shall be visual characters complying with Section 703.2.4.

EXCEPTION: Bus schedules, timetables and maps that are posted at the bus stop or bus bay shall not be required to comply with Section 805.4.

805.5 Rail Platforms. Rail platforms shall comply with Section 805.5.

805.5.1 Slope. Rail platforms shall not exceed a slope of 1:48 in all directions.

EXCEPTION: Where platforms serve vehicles operating on existing track or track laid in existing roadway, the slope of the platform parallel to the track shall be permitted to be equal to the slope (grade) of the roadway or existing track.

805.5.2 Detectable Warnings. Platform boarding edges not protected by platform screens or guards shall have a detectable warning complying with Section 705.

805.6 Rail Station Signs. Rail station signs shall comply with Section 805.6.

EXCEPTION: Signs shall not be required to comply with Sections 805.6.1 and 805.6.2 where audible signs are remotely transmitted to hand-held receivers, or are user- or proximity-actuated.

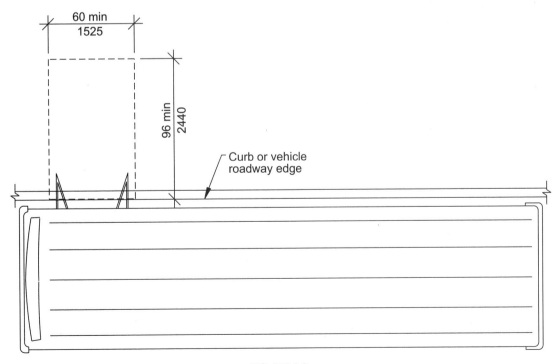

FIG. 805.2.2
SIZE OF BUS BOARDING AND ALIGHTING AREAS

805.6.1 Entrances. Where signs identify a station or a station entrance, at least one sign with raised characters and braille complying with Sections 703.3 and 703.4 shall be provided at each entrance.

805.6.2 Routes and Destinations. Lists of stations, routes and destinations served by the station that are located on boarding areas, platforms, or mezzanines shall have visual characters complying with Section 703.2. A minimum of one sign with raised characters and braille complying with Sections 703.3 and 703.4 shall be provided on each platform or boarding area to identify the specific station.

> **EXCEPTION:** Where sign space is limited, characters shall not be required to exceed 3 inches (75 mm) in height.

805.6.3 Station Names. Stations covered by this section shall have identification signs with visual characters complying with Section 703.2. The signs shall be clearly visible and within the sight lines of a standing or sitting passenger from within the vehicle on both sides when not obstructed by another vehicle.

805.7 Public Address Systems. Where public address systems convey audible information to the public, the same or equivalent information shall be provided in a visual format.

805.8 Clocks. Where clocks are provided for use by the public, the clock face shall be uncluttered so that its elements are clearly visible. Hands, numerals and digits shall contrast with the background either light-on-dark or dark-on-light. Where clocks are installed overhead, numerals and digits shall be visual characters complying with Section 703.2.

805.9 Escalators. Where provided, escalators shall have a 32-inch (815mm) minimum clear width, and shall comply with Requirements 6.1.3.5.6-Step Demarcations, and 6.1.3.6.5-Flat Steps of ASME A17.1/CSA B44 listed in Section 105.2.5.

> **EXCEPTION:** Existing escalators shall not be required to comply with Section 805.9.

805.10 Track Crossings. Where a circulation path crosses tracks, it shall comply with Section 402 and shall have a detectable warning 24 inches (610 mm) in depth complying with Section 705 extending the full width of the circulation path. The detectable warning

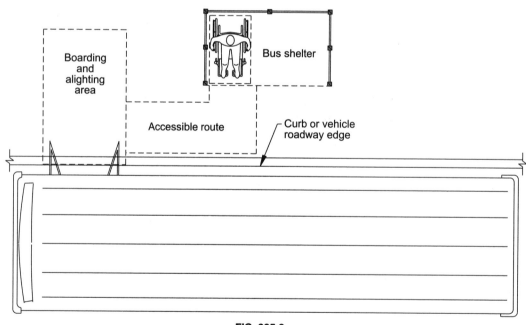

FIG. 805.3
BUS SHELTERS

FIG. 805.10
TRACK CROSSINGS

surface shall be located so that the edge nearest the rail crossing is 6 foot (1830 mm) minimum and 15 foot (4570 mm) maximum from the centerline of the nearest rail.

EXCEPTION: Openings for wheel flanges shall be permitted to be $2^1/_2$ inches (64 mm) maximum.

806 Holding Cells and Housing Cells

806.1 General. Holding cells and housing cells shall comply with Section 806.

806.2 Features for People Using Wheelchairs or Other Mobility Aids. Cells required to have features for people using wheelchairs or other mobility aids shall comply with Section 806.2.

806.2.1 Turning Space. Turning space complying with Section 304 shall be provided within the cell.

806.2.2 Benches. Where benches are provided, at least one bench shall comply with Section 903.

806.2.3 Beds. Where beds are provided, clear floor space complying with Section 305 shall be provided on at least one side of the bed. The clear floor space shall be positioned for parallel approach to the side of the bed.

806.2.4 Toilet and Bathing Facilities. Toilet facilities or bathing facilities provided as part of a cell shall comply with Section 603.

806.3 Communication Features. Cells required to have communication features shall comply with Section 806.3.

806.3.1 Alarms. Where audible emergency alarm systems are provided to serve the occupants of cells, visible alarms complying with Section 702 shall be provided.

EXCEPTION: In cells where inmates or detainees are not allowed independent means of egress, visible alarms shall not be required.

806.3.2 Telephones. Where provided, telephones within cells shall have volume controls complying with Section 704.3.

807 Courtrooms

807.1 General. Courtrooms shall comply with Section 807.

807.2 Turning Space. Where provided, each area that is raised or depressed shall provide a turning space complying with Section 304.

EXCEPTION: Levels of jury boxes not required to be accessible are not required to comply with Section 807.2.

807.3 Clear Floor Space. Within the defined area of each jury box and witness stand, a clear floor space complying with Section 305 shall be provided.

EXCEPTION: In alterations, wheelchair spaces are not required to be located within the defined area of raised jury boxes or witness stands and shall be per-mitted to be located outside these spaces where ramps or platform lifts restrict or project into the means of egress required by the administrative authority.

807.4 Courtroom Stations. Judges' benches, clerks' stations, bailiffs' stations, deputy clerks' stations, court reporters' stations and litigants' and counsel stations shall comply with Section 902.

807.5 Gallery seating. Gallery seating shall comply with Section 802.

Chapter 9. Built-in Furnishings and Equipment

901 General

901.1 Scope. Built-in furnishings and equipment required to be accessible by the scoping provisions adopted by the administrative authority shall comply with the applicable provisions of Chapter 9.

902 Dining Surfaces and Work Surfaces

902.1 General. Accessible dining surfaces and work surfaces shall comply with Section 902.

> **EXCEPTION:** Dining surfaces and work surfaces primarily for children's use shall be permitted to comply with Section 902.5.

902.2 Clear Floor Space. A clear floor space complying with Section 305, positioned for a forward approach, shall be provided. Knee and toe clearance complying with Section 306 shall be provided.

> **EXCEPTIONS:**
>
> 1. At drink surfaces 12 inches (305 mm) or less in depth, knee and toe space shall not be required to extend beneath the surface beyond the depth of the drink surface provided.
>
> 2. Dining surfaces that are 15 inches (380 mm) minimum and 24 inches (610 mm) maximum in height are permitted to have a clear floor space complying with Section 305 positioned for a parallel approach.

902.3 Exposed Surfaces. There shall be no sharp or abrasive surfaces under the exposed portions of dining surfaces and work surfaces.

902.4 Height. The tops of dining surfaces and work surfaces shall be 28 inches (710 mm) minimum and 34 inches (865 mm) maximum in height above the floor.

902.5 Dining Surfaces and Work Surfaces for Children's Use. Accessible dining surfaces and work surfaces primarily for children's use shall comply with Section 902.5.

> **EXCEPTION:** Dining surfaces and work surfaces used primarily by children ages 5 and younger shall not be required to comply with Section 902.5 where a clear floor space complying with Section 305 is provided and is positioned for a parallel approach.

902.5.1 Clear Floor Space. A clear floor space complying with Section 305, positioned for forward approach, shall be provided. Knee and toe clearance complying with Section 306 shall be provided.

> **EXCEPTION:** A knee clearance of 24 inches (610 mm) minimum above the floor shall be permitted.

902.5.2 Height. The tops of tables and counters shall be 26 inches (660 mm) minimum and 30 inches (760 mm) maximum above the floor.

903 Benches

903.1 General. Accessible benches shall comply with Section 903.

903.2 Clear Floor Space. A clear floor space complying with Section 305, positioned for parallel approach to the bench seat, shall be provided.

903.3 Size. Benches shall have seats 42 inches (1065 mm) minimum in length, and 20 inches (510 mm) minimum and 24 inches (610 mm) maximum in depth.

903.4 Back Support. The bench shall provide for back support or shall be affixed to a wall. Back support shall be 42 inches (1065 mm) minimum in length and shall extend from a point 2 inches (51 mm) maximum above the seat surface to a point 18 inches (455 mm) minimum above the seat surface. Back support shall be $2^1/_2$ inches (64 mm) maximum from the rear edge of the seat measured horizontally.

903.5 Height. The top of the bench seat shall be 17 inches (430 mm) minimum and 19 inches (485 mm) maximum above the floor, measured to the top of the seat.

> **EXCEPTION:** Benches primarily for children's use shall be permitted to be 11 inches (280 mm) minimum and 17 inches (430 mm) maximum above the floor, measured to the top of the seat.

903.6 Structural Strength. Allowable stresses shall not be exceeded for materials used where a vertical or horizontal force of 250 pounds (1112 N) is applied at any point on the seat, fastener mounting device, or supporting structure.

903.7 Wet Locations. Where provided in wet locations the surface of the seat shall be slip resistant and shall not accumulate water.

904 Sales and Service Counters

904.1 General. Accessible sales and service counters and windows shall comply with Section 904 as applicable.

> **EXCEPTION:** Drive up only sales or service counters and windows are not required to comply with Section 904.

904.2 Approach. All portions of counters required to be accessible shall be located adjacent to a walking surface complying with Section 403.

904.3 Sales and Service Counters. Sales and service counters shall comply with Section 904.3.1 or 904.3.2. The accessible portion of the countertop shall extend the same depth as the sales and service countertop.

904.3.1 Parallel Approach. A portion of the counter surface 36 inches (915 mm) minimum in length and 36 inches (915 mm) maximum in height above the floor shall be provided. Where the counter surface is less than 36 inches (915 mm) in length, the entire counter surface shall be 36 inches (915 mm) maximum in height above the floor. A clear floor space complying with Section 305, positioned for a parallel approach adjacent to the accessible counter, shall be provided.

904.3.2 Forward Approach. A portion of the counter surface 30 inches (760 mm) minimum in length and 36 inches (915 mm) maximum in height above the floor shall be provided. A clear floor space complying with Section 305, positioned for a forward approach to the accessible counter, shall be provided. Knee and toe clearance complying with Section 306 shall be provided under the accessible counter.

904.4 Checkout Aisles. Checkout aisles shall comply with Section 904.4.

904.4.1 Aisle. Aisles shall comply with Section 403.

904.4.2 Counters. The checkout counter surface shall be 38 inches (965 mm) maximum in height above the floor. The top of the counter edge protection shall be 2 inches (51 mm) maximum above the top of the counter surface on the aisle side of the checkout counter.

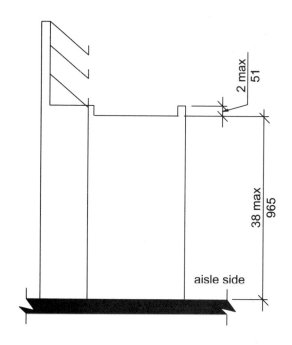

FIG. 904.4.2
HEIGHT OF CHECKOUT COUNTERS

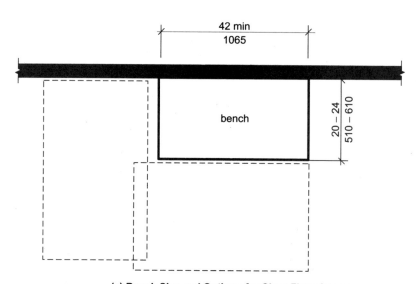

(a) Bench Size and Options for Clear Floor Space

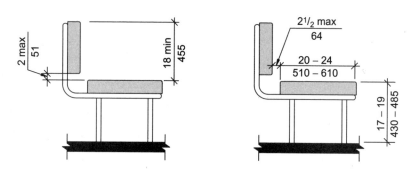

(b) Bench Back Support and Seat Height

FIG. 903
BENCHES

904.4.3 Check Writing Surfaces. Where provided, check writing surfaces shall comply with Section 902.4.

904.5 Food Service Lines. Counters in food service lines shall comply with Section 904.5.

904.5.1 Self-Service Shelves and Dispensing Devices. Self-service shelves and dispensing devices for tableware, dishware, condiments, food and beverages shall comply with Section 308.

904.5.2 Tray Slides. The tops of tray slides shall be 28 inches (710 mm) minimum and 34 inches (865 mm) maximum above the floor.

904.6 Security Glazing. Where counters or teller windows have security glazing to separate personnel from the public, a method to facilitate voice communication shall be provided. Telephone handset devices, if provided, shall comply with Section 704.3.

905 Storage Facilities

905.1 General. Accessible storage facilities shall comply with Section 905.

905.2 Clear Floor Space. A clear floor space complying with Section 305 shall be provided.

905.3 Height. Accessible storage elements shall comply with at least one of the reach ranges specified in Section 308.

905.4 Operable Parts. Operable parts of storage facilities shall comply with Section 309.

Chapter 10. Dwelling Units and Sleeping Units

1001 General

1001.1 Scoping. Dwelling units and sleeping units required to be Accessible units, Type A units, Type B units, Type C (Visitable) units or units with accessible communication features by the scoping provisions adopted by the administrative authority shall comply with the applicable provisions of Chapter 10.

1002 Accessible Units

1002.1 General. Accessible units shall comply with Section 1002.

1002.2 Primary Entrance. The accessible primary entrance shall be on an accessible route from public and common areas. The primary entrance shall not be to a bedroom unless it is the only entrance.

1002.3 Accessible Route. Accessible routes within Accessible units shall comply with Section 1002.3.

1002.3.1 Location. At least one accessible route shall connect all spaces and elements that are a part of the unit. Accessible routes shall coincide with or be located in the same area as a general circulation path.

EXCEPTION: An accessible route is not required to unfinished attics and unfinished basements that are part of the unit.

1002.3.2 Turning Space. All rooms served by an accessible route shall provide a turning space complying with Section 304.

EXCEPTIONS:

1. A turning space shall not be required in toilet rooms and bathrooms that are not required to comply with Section 1002.11.2.

2. A turning space is not required within closets or pantries that are 48 inches (1220 mm) maximum in depth.

1002.3.3 Components. Accessible routes shall consist of one or more of the following elements: walking surfaces with a slope not steeper than 1:20, doors and doorways, ramps, elevators, and platform lifts.

1002.4 Walking Surfaces. Walking surfaces that are part of an accessible route shall comply with Section 403.

1002.5 Doors and Doorways. The primary entrance door to the unit, and all other doorways intended for user passage, shall comply with Section 404.

EXCEPTIONS:

1. Existing doors to hospital patient sleeping rooms shall be exempt from the requirement for space at the latch side provided the door is 44 inches (1120 mm) minimum in width.

2. In toilet rooms and bathrooms not required to comply with Section 1002.11.2, maneuvering clearances required by Section 404.2.3 are not required on the toilet room or bathroom side of the door.

3. A turning space between doors in a series as required by Section 404.2.5 is not required.

4. Storm and screen doors are not required to comply with Section 404.2.5.

5. Communicating doors between individual sleeping units are not required to comply with Section 404.2.5.

6. At other than the primary entrance door, where exterior space dimensions of balconies are less than the required maneuvering clearance, door maneuvering clearance is not required on the exterior side of the door.

1002.6 Ramps. Ramps shall comply with Section 405.

1002.7 Elevators. Elevators within the unit shall comply with Section 407, 408, or 409.

1002.8 Platform Lifts. Platform lifts within the unit shall comply with Section 410.

1002.9 Operable Parts. Lighting controls, electrical panelboards, electrical switches and receptacle outlets, environmental controls, appliance controls, operating hardware for operable windows, plumbing fixture controls, and user controls for security or intercom systems shall comply with Section 309.

EXCEPTIONS:

1. Receptacle outlets serving a dedicated use.

2. Where two or more receptacle outlets are provided in a kitchen above a length of counter top that is uninterrupted by a sink or appliance, one receptacle outlet shall not be required to comply with 309.

3. Floor receptacle outlets.

4. HVAC diffusers.

5. Controls mounted on ceiling fans.

6. Where redundant controls other than light switches are provided for a single element, one control in each space shall not be required to be accessible.

7. Reset buttons and shut-offs serving appliances, piping and plumbing fixtures.

8. Electrical panelboards shall not be required to comply with Section 309.4.

1002.10 Laundry Equipment. Washing machines and clothes dryers shall comply with Section 611.

1002.11 Toilet and Bathing Facilities. At least one toilet and bathing facility shall comply with Section

1002.11.2. All other toilet and bathing facilities shall comply with Section 1002.11.1

1002.11.1 Grab Bars and Shower Seat Reinforcement. At fixtures in toilet and bathing facilities not required to comply with Section 1002.11.2, reinforcement in accordance with Section 1004.11.1 shall be provided.

EXCEPTION: Reinforcement is not required where Type B units are not provided in the structure.

1002.11.2 Accessible Toilet and Bathing Facility. At least one toilet and bathing facility shall comply with Section 603. At least one lavatory, one water closet and either a bathtub or shower within the unit shall comply with Sections 604 through 610. The accessible toilet and bathing fixtures shall be in a single toilet/bathing area, such that travel between fixtures does not require travel through other parts of the unit.

1002.11.2.1 Vanity Counter Top Space. If vanity counter top space is provided in dwelling or sleeping units not required to be Accessible units within the same facility, equivalent vanity counter top space, in terms of size and proximity to the lavatory, shall also be provided in Accessible units.

1002.11.2.2 Mirrors. Mirrors above accessible lavatories shall have the bottom edge of the reflecting surface 40 inches (1015 mm) maximum above the floor.

1002.12 Kitchens and kitchenettes. Kitchens and kitchenettes shall comply with Section 804. At least one work surface, 30 inches (760 mm) minimum in length, shall comply with Section 902.

EXCEPTION: Spaces that do not provide a cooktop or conventional range shall not be required to provide an accessible work surface.

1002.13 Windows. Windows shall comply with Section 1002.13.

1002.13.1 Natural ventilation. Operable windows required to provide natural ventilation shall comply with Sections 309.2 and 309.3.

1002.13.2 Emergency escape. Operable windows required to provide an emergency escape and rescue opening shall comply with Section 309.2.

1002.14 Storage Facilities. Where storage facilities are provided, at least one of each type shall comply with Section 905.

EXCEPTION: Kitchen cabinets shall not be required to comply with Section 1002.14.

1002.15 Beds. In at least one sleeping area, a minimum of five percent, but not less than one bed shall comply with Section 1002.15.

1002.15.1 Clear Floor Space. A clear floor space complying with Section 305 shall be provided on both sides of the bed. The clear floor space shall be positioned for parallel approach to the side of the bed.

EXCEPTION: Where a single clear floor space complying with Section 305 positioned for parallel approach is provided between two beds, a clear floor space shall not be required on both sides of the bed.

1002.15.2 Bed Frames. At least one bed shall be provided with an open bed frame.

1003 Type A Units

1003.1 General. Type A units shall comply with Section 1003.

1003.2 Primary Entrance. The accessible primary entrance shall be on an accessible route from public and common areas. The primary entrance shall not be to a bedroom unless it is the only entrance.

1003.3 Accessible Route. Accessible routes within Type A units shall comply with Section 1003.3.

1003.3.1 Location. At least one accessible route shall connect all spaces and elements that are a part of the unit. Accessible routes shall coincide with or be located in the same area as a general circulation path.

EXCEPTION: An accessible route is not required to unfinished attics and unfinished basements that are part of the unit.

1003.3.2 Turning Space. All rooms served by an accessible route shall provide a turning space complying with Section 304.

EXCEPTIONS:

1. A turning space is not required in toilet rooms and bathrooms that are not required to comply with Section 1003.11.2.

2. A turning space is not required within closets or pantries that are 48 inches (1220 mm) maximum in depth.

1003.3.3 Components. Accessible routes shall consist of one or more of the following elements: walking surfaces with a slope not steeper than 1:20, doors and doorways, ramps, elevators, and platform lifts.

1003.4 Walking Surfaces. Walking surfaces that are part of an accessible route shall comply with Section 403.

1003.5 Doors and Doorways. The primary entrance door to the unit, and all other doorways intended for user passage, shall comply with Section 404.

EXCEPTIONS:

1. Thresholds at exterior sliding doors shall be permitted to be $^3/_4$ inch (19 mm) maximum in height, provided they are beveled with a slope not greater than 1:2.

2. In toilet rooms and bathrooms not required to comply with Section 1003.11.2, maneuvering clearances required by Section 404.2.3 are not required on the toilet room or bathroom side of the door.

3. A turning space between doors in a series as required by Section 404.2.5 is not required.

4. Storm and screen doors are not required to comply with Section 404.2.5.

5. Communicating doors between individual sleeping units are not required to comply with Section 404.2.5.

6. At other than the primary entrance door, where exterior space dimensions of balconies are less than the required maneuvering clearance, door maneuvering clearance is not required on the exterior side of the door.

1003.6 Ramps. Ramps shall comply with Section 405.

1003.7 Elevators. Elevators within the unit shall comply with Section 407, 408, or 409.

1003.8 Platform Lifts. Platform lifts within the unit shall comply with Section 410.

1003.9 Operable Parts. Lighting controls, electrical panelboards, electrical switches and receptacle outlets, environmental controls, appliance controls, operating hardware for operable windows, plumbing fixture controls, and user controls for security or intercom systems shall comply with Section 309.

EXCEPTIONS:

1. Receptacle outlets serving a dedicated use.

2. Where two or more receptacle outlets are provided in a kitchen above a length of counter top that is uninterrupted by a sink or appliance, one receptacle outlet shall not be required to comply with Section 309.

3. Floor receptacle outlets.

4. HVAC diffusers.

5. Controls mounted on ceiling fans.

6. Where redundant controls other than light switches are provided for a single element, one control in each space shall not be required to be accessible.

7. Reset buttons and shut-offs serving appliances, piping and plumbing fixtures.

8. Electrical panelboards shall not be required to comply with Section 309.4.

1003.10 Laundry Equipment. Washing machines and clothes dryers shall comply with Section 611.

1003.11 Toilet and Bathing Facilities. At least one toilet and bathing facility shall comply with Section 1003.11.2. All toilet and bathing facilities shall comply with Section 1003.11.1.

1003.11.1 Grab Bar and Shower Seat Reinforcement. Reinforcement shall be provided for the future installation of grab bars complying with Section 604.5 at water closets; grab bars complying with Section 607.4 at bathtubs; and for grab bars and shower seats complying with Sections 608.3, 608.2.1.3, 608.2.2.3 and 608.2.3.2 at shower compartments.

EXCEPTIONS:

1. At fixtures not required to comply with Section 1003.11.2, reinforcement in accordance with Section 1004.11.1 shall be permitted.

2. Reinforcement is not required in a room containing only a lavatory and a water closet, provided the room does not contain the only lavatory or water closet on the accessible level of the dwelling unit.

3. Reinforcement for the water closet side wall vertical grab bar component required by Section 604.5 is not required.

4. Where the lavatory overlaps the water closet clearance in accordance with the exception to Section 1003.11.2.4.4 reinforcement at the water closet rear wall for a 24-inch (610 mm) minimum length grab bar, centered on the water closet, shall be provided.

1003.11.2 General. At least one toilet and bathing facility shall comply with Section 1003.11.2. At least one lavatory, one water closet and either a bathtub or shower within the unit shall comply with Section 1003.11.2. The accessible toilet and bathing fixtures shall be in a single toilet/bathing area, such that travel between fixtures does not require travel through other parts of the unit.

1003.11.2.1 Doors. Doors shall not swing into the clear floor space or clearance for any fixture.

EXCEPTION: Where a clear floor space complying with Section 305.3 is provided within the room beyond the arc of the door swing.

1003.11.2.2 Lavatory. Lavatories shall comply with Section 606.

EXCEPTION: Cabinetry shall be permitted under the lavatory, provided the following criteria are met:

(a) The cabinetry can be removed without removal or replacement of the lavatory;

(b) The floor finish extends under the cabinetry; and

(c) The walls behind and surrounding the cabinetry are finished.

1003.11.2.3 Mirrors. Mirrors above accessible lavatories shall have the bottom edge of the reflecting surface 40 inches (1015 mm) maximum above the floor.

1003.11.2.4 Water Closet. Water closets shall comply with Section 1003.11.2.4.

1003.11.2.4.1 Location. The water closet shall be positioned with a wall to the rear and to one side. The centerline of the water closet shall be

16 inches (405 mm) minimum and 18 inches (455 mm) maximum from the sidewall.

1003.11.2.4.2 Clearance Width. Clearance around the water closet shall be 60 inches (1525 mm) minimum in width, measured perpendicular from the side wall.

1003.11.2.4.3 Clearance Depth. Clearance around the water closet shall be 56 inches (1420 mm) minimum in depth, measured perpendicular from the rear wall.

1003.11.2.4.4 Clearance Overlap. The required clearance around the water closet shall be permitted to overlap the water closet, associated grab bars, paper dispensers, coat hooks, shelves, accessible routes, clear floor space required at other fixtures, and the wheelchair turning space. No other fixtures or obstructions shall be located within the required water closet clearance.

EXCEPTION: A lavatory measuring 24 inches (610 mm) maximum in depth and complying with Section 1003.11.2.2 shall be permitted on the rear wall 18 inches (455 mm) minimum from the centerline of the water closet to the side edge of the lavatory where the clearance at the water closet is 66 inches (1675 mm) minimum measured perpendicular from the rear wall.

1003.11.2.4.5 Height. The top of the water closet seat shall be 15 inches (380 mm) minimum and 19 inches (485 mm) maximum above the floor, measured to the top of the seat.

1003.11.2.4.6 Flush Controls. Flush controls shall be hand-operated or automatic. Hand operated flush controls shall comply with Section 309. Hand-operated flush controls shall be located on the open side of the water closet.

1003.11.2.5 Bathing Fixtures. The accessible bathing fixture shall be a bathtub complying with Section 1003.11.2.5.1 or a shower compartment complying with Section 1003.11.2.5.2.

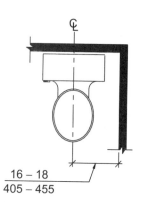

(a) Water Closet Location

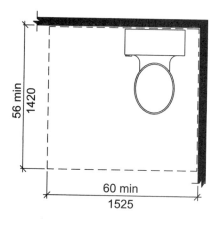

(b) Minimum Clearance

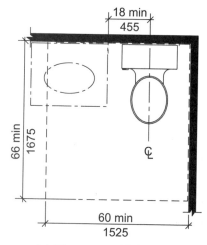

**(c) Clearance with Lavatory
(Overlap Exception)**

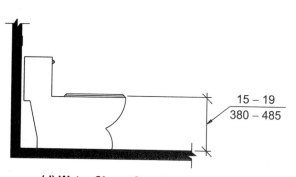

(d) Water Closet Seat Height

**FIG. 1003.11.2.4
WATER CLOSETS IN TYPE A UNITS**

1003.11.2.5.1 Bathtub. Bathtubs shall comply with Section 607.

EXCEPTIONS:

1. The removable in-tub seat required by Section 607.3 is not required.

2. Counter tops and cabinetry shall be permitted at one end of the clearance, provided the following criteria are met:

    (a) The countertop and cabinetry can be removed;

    (b) The floor finish extends under the countertop and cabinetry; and

    (c) The walls behind and surrounding the countertop and cabinetry are finished.

1003.11.2.5.2 Shower. Showers shall comply with Section 608.

EXCEPTION: At standard roll-in shower compartments complying with Section 608.2.2, lavatories, counter tops and cabinetry shall be permitted at one end of the clearance, provided the following criteria are met:

    (a) The countertop and cabinetry can be removed;

    (b) The floor finish extends under the countertop and cabinetry; and

    (c) The walls behind and surrounding the countertop and cabinetry are finished.

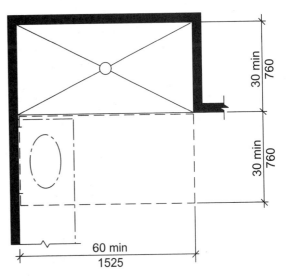

Note: Lavatory permitted per Section 608.2.2

FIG. 1003.11.2.5.2
STANDARD ROLL-IN-TYPE SHOWER
COMPARTMENT IN TYPE A UNITS

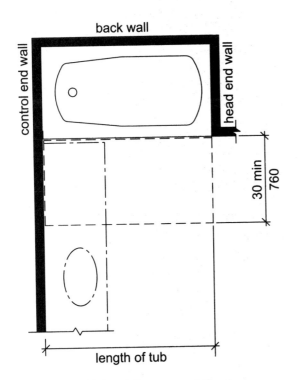

(a) Without Permanent Seat

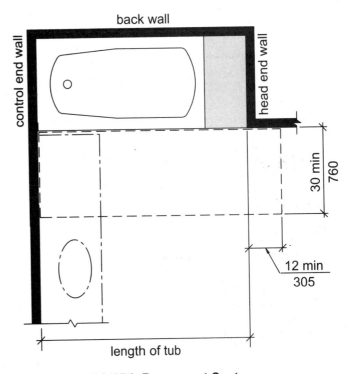

(b) With Permanent Seat

FIG. 1003.11.2.5.1
CLEARANCE FOR BATHTUBS IN TYPE A UNITS

1003.12 Kitchens and kitchenettes. Kitchens and kitchenettes shall comply with Section 1003.12.

1003.12.1 Clearance. Clearance complying with Section 1003.12.1 shall be provided.

1003.12.1.1 Minimum Clearance. Clearance between all opposing base cabinets, counter tops, appliances, or walls within kitchen work areas shall be 40 inches (1015mm) minimum.

1003.12.1.2 U-Shaped Kitchens. In kitchens with counters, appliances, or cabinets on three contiguous sides, clearance between all opposing base cabinets, countertops, appliances, or walls within kitchen work areas shall be 60 inches (1525 mm) minimum.

1003.12.2 Clear Floor Space. Clear floor spaces required by Sections 1003.12.3 through 1003.12.5 shall comply with Section 305.

1003.12.3 Work Surface. At least one section of counter shall provide a work surface 30 inches (760 mm) minimum in length complying with Section 1003.12.3.

1003.12.3.1 Clear Floor Space. A clear floor space, positioned for a forward approach to the work surface, shall be provided. Knee and toe clearance complying with Section 306 shall be provided. The clear floor space shall be centered on the work surface.

EXCEPTION: Cabinetry shall be permitted under the work surface, provided the following criteria are met:

(a) The cabinetry can be removed without removal or replacement of the work surface,

(b) The floor finish extends under the cabinetry, and

(c) The walls behind and surrounding the cabinetry are finished.

1003.12.3.2 Height. The work surface shall be 34 inches (865 mm) maximum above the floor.

EXCEPTION: A counter that is adjustable to provide a work surface at variable heights 29 inches (735 mm) minimum and 36 inches (915 mm) maximum above the floor, or that can be relocated within that range without cutting the counter or damaging adjacent cabinets, walls, doors, and structural elements, shall be permitted.

1003.12.3.3 Exposed Surfaces. There shall be no sharp or abrasive surfaces under the exposed portions of work surface counters.

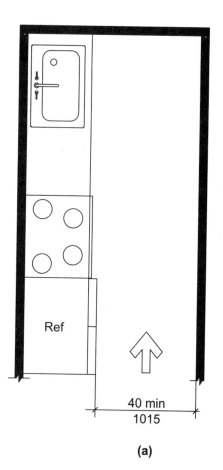

(a)

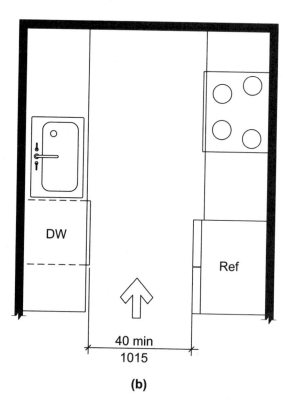

(b)

FIG. 1003.12.1.1
MINIMUM KITCHEN CLEARANCE IN TYPE A UNITS

1003.12.4 Sink. The sink shall comply with Section 1003.12.4.

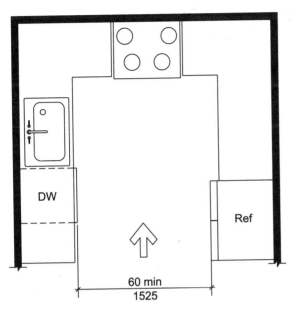

FIG. 1003.12.1.2
U-SHAPED KITCHEN CLEARANCE IN TYPE A UNITS

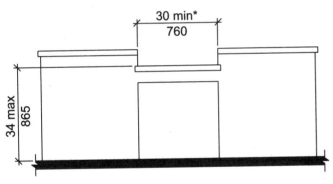

* 36 min. (915) if part of T-shaped turning space per Sections 304.3.2 and 1003.3.2

FIG. 1003.12.3
WORK SURFACE IN KITCHEN FOR TYPE A UNITS

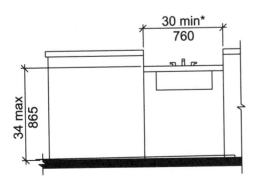

* 36 min. (915) if part of T-shaped turning space per Sections 304.3.2 and 1003.3.2

FIG. 1003.12.4
KITCHEN SINK FOR TYPE A UNITS

1003.12.4.1 Clear Floor Space. A clear floor space, positioned for a forward approach to the sink, shall be provided. Knee and toe clearance complying with Section 306 shall be provided.

EXCEPTIONS:

1. The requirement for knee and toe clearance shall not apply to more than one bowl of a multi-bowl sink.

2. Cabinetry shall be permitted to be added under the sink, provided the following criteria are met:

    (a) The cabinetry can be removed without removal or replacement of the sink,

    (b) The floor finish extends under the cabinetry, and

    (c) The walls behind and surrounding the cabinetry are finished.

1003.12.4.2 Height. The front of the sink shall be 34 inches (865 mm) maximum above the floor, measured to the higher of the rim or counter surface.

EXCEPTION: A sink and counter that is adjustable to variable heights 29 inches (735 mm) minimum and 36 inches (915 mm) maximum above the floor, or that can be relocated within that range without cutting the counter or damaging adjacent cabinets, walls, doors and structural elements, provided rough-in plumbing permits connections of supply and drain pipes for sinks mounted at the height of 29 inches (735 mm), shall be permitted.

1003.12.4.3 Faucets. Faucets shall comply with Section 309.

1003.12.4.4 Exposed Pipes and Surfaces. Water supply and drain pipes under sinks shall be insulated or otherwise configured to protect against contact. There shall be no sharp or abrasive surfaces under sinks.

1003.12.5 Appliances. Where provided, kitchen appliances shall comply with Section 1003.12.5.

1003.12.5.1 Operable Parts. All appliance controls shall comply with Section 1003.9.

EXCEPTIONS:

1. Appliance doors and door latching devices shall not be required to comply with Section 309.4.

2. Bottom-hinged appliance doors, when in the open position, shall not be required to comply with Section 309.3.

1003.12.5.2 Clear Floor Space. A clear floor space, positioned for a parallel or forward approach, shall be provided at each kitchen appliance.

1003.12.5.3 Dishwasher. A clear floor space, positioned adjacent to the dishwasher door, shall be provided. The dishwasher door in the open position shall not obstruct the clear floor space for the dishwasher or an adjacent sink.

1003.12.5.4 Cooktop. Cooktops shall comply with Section 1003.12.5.4.

1003.12.5.4.1 Approach. A clear floor space, positioned for a parallel or forward approach to the cooktop, shall be provided.

1003.12.5.4.2 Forward approach. Where the clear floor space is positioned for a forward approach, knee and toe clearance complying with Section 306 shall be provided. The underside of the cooktop shall be insulated or otherwise configured to protect from burns, abrasions, or electrical shock.

1003.12.5.4.3 Parallel approach. Where the clear floor space is positioned for a parallel approach, the clear floor space shall be centered on the appliance.

1003.12.5.4.4 Controls. The location of controls shall not require reaching across burners.

1003.12.5.5 Oven. Ovens shall comply with Section 1003.12.5.5. Ovens shall have controls on front panels, on either side of the door.

1003.12.5.5.1 Clear floor space. A clear floor space shall be provided. The oven door in the open position shall not obstruct the clear floor space for the oven.

1003.12.5.5.2 Side-Hinged Door Ovens. Side-hinged door ovens shall have a countertop positioned adjacent to the latch side of the oven door.

1003.12.5.5.3 Bottom-Hinged Door Ovens. Bottom-hinged door ovens shall have a countertop positioned adjacent to one side of the door.

1003.12.5.5.4 Controls. The location of controls shall not require reaching across burners.

1003.12.5.6 Refrigerator/Freezer. Combination refrigerators and freezers shall have at least 50 percent of the freezer compartment shelves, including the bottom of the freezer 54 inches (1370 mm) maximum above the floor when the shelves are installed at the maximum heights possible in the compartment. A clear floor space, positioned for a parallel approach to the refrigerator/freezer, shall be provided. The centerline of the clear floor space shall be offset 24 inches (610 mm) maximum from the centerline of the appliance.

1003.13 Windows. Windows shall comply with Section 1003.13.

1003.13.1 Natural ventilation. Operable windows required to provide natural ventilation shall comply with Sections 309.2 and 309.3.

1003.13.2 Emergency escape. Operable windows required to provide an emergency escape and rescue opening shall comply with Section 309.2.

1003.14 Storage Facilities. Where storage facilities are provided, at least one of each type shall comply with Section 905.

EXCEPTION: Kitchen cabinets shall not be required to comply with Section 1003.14.

1004 Type B Units

1004.1 General. Type B units shall comply with Section 1004.

1004.2 Primary Entrance. The accessible primary entrance shall be on an accessible route from public and common areas. The primary entrance shall not be to a bedroom unless it is the only entrance.

1004.3 Accessible Route. Accessible routes within Type B units shall comply with Section 1004.3.

1004.3.1 Location. At least one accessible route shall connect all spaces and elements that are a part of the unit. Accessible routes shall coincide with or be located in the same area as a general circulation path.

EXCEPTIONS:

1. An accessible route is not required to unfinished attics and unfinished basements that are part of the unit.

2. One of the following is not required to be on an accessible route:

   2.1 A raised floor area in a portion of a living, dining, or sleeping room; or

   2.2 A sunken floor area in a portion of a living, dining, or sleeping room; or

   2.3 A mezzanine that does not have plumbing fixtures or an enclosed habitable space.

1004.3.2 Components. Accessible routes shall consist of one or more of the following elements: walking surfaces with a slope not steeper than 1:20, doors and doorways, ramps, elevators, and platform lifts.

1004.4 Walking Surfaces. Walking surfaces that are part of an accessible route shall comply with Section 1004.4.

1004.4.1 Clear Width. Clear width of an accessible route shall comply with Section 403.5.

1004.4.2 Changes in Level. Changes in level shall comply with Section 303.

EXCEPTION: Where exterior deck, patio or balcony surface materials are impervious, the finished exterior impervious surface shall be 4 inches (100 mm) maximum below the floor level of the adjacent interior spaces of the unit.

1004.5 Doors and Doorways. Doors and doorways shall comply with Section 1004.5.

1004.5.1 Primary Entrance Door. The primary entrance door to the unit shall comply with Section 404.

EXCEPTION: Storm and screen doors serving individual dwelling or sleeping units are not required to comply with Section 404.2.5.

1004.5.2 User Passage Doorways. Doorways intended for user passage shall comply with Section 1004.5.2.

1004.5.2.1 Clear Width. Doorways shall have a clear opening of $31^3/_4$ inches (805 mm) minimum. Clear opening of swinging doors shall be measured between the face of the door and stop, with the door open 90 degrees.

1004.5.2.1.1 Double Leaf Doorways. Where the operable parts on an inactive leaf of a double leaf doorway are located more than 48 inches (1220 mm) or less than 15 inches (380 mm) above the floor, the active leaf shall provide the clearance required by Section 1004.5.2.1.

1004.5.2.2 Thresholds. Thresholds shall comply with Section 303.

EXCEPTION: Thresholds at exterior slidng doors shall be permitted to be $^3/_4$ inch (19 mm) maximum in height, provided they are beveled with a slope not steeper than 1:2.

1004.5.2.3 Automatic Doors. Automatic doors shall comply with Section 404.3.

1004.6 Ramps. Ramps shall comply with Section 405.

1004.7 Elevators. Elevators within the unit shall comply with Section 407, 408, or 409.

1004.8 Platform Lifts. Platform lifts within the unit shall comply with Section 410.

1004.9 Operable Parts. Lighting controls, electrical switches and receptacle outlets, environmental controls, electrical panelboards, and user controls for security or intercom systems shall comply with Sections 309.2 and 309.3.

EXCEPTIONS:

1. Receptacle outlets serving a dedicated use.

2. Where two or more receptacle outlets are provided in a kitchen above a length of counter top that is uninterrupted by a sink or appliance, one receptacle outlet shall not be required to comply with Section 309.

3. Floor receptacle outlets.

4. HVAC diffusers.

5. Controls mounted on ceiling fans.

6. Controls or switches mounted on appliances.

7. Plumbing fixture controls.

8. Reset buttons and shut-offs serving appliances, piping and plumbing fixtures.

9. Where redundant controls other than light switches are provided for a single element, one control in each space shall not be required to be accessible.

10. Within kitchens and bathrooms, lighting controls, electrical switches and receptacle outlets are permitted to be located over cabinets with counter tops 36 inches (915 mm) maximum in height and 25 $^1/_2$ inches (650 mm) maximum in depth.

1004.10 Laundry Equipment. Washing machines and clothes dryers shall comply with Section 1004.10.

1004.10.1 Clear Floor Space. A clear floor space complying with Section 305.3, shall be provided. A parallel approach shall be provided for a top loading machine. A forward or parallel approach shall be provided for a front loading machine.

1004.11 Toilet and Bathing Facilities. Toilet and bathing fixtures shall comply with Section 1004.11.

EXCEPTION: Fixtures on levels not required to be accessible.

1004.11.1 Grab Bar and Shower Seat Reinforcement. Reinforcement shall be provided for the future installation of grab bars and shower seats at water closets, bathtubs, and shower compartments. Where walls are located to permit the installation of grab bars and seats complying with Section 604.5 at water closets; grab bars complying with Section 607.4 at bathtubs; and for grab bars and shower seats complying with Sections, 608.3, 608.2.1.3, 608.2.2.3 and 608.2.3.2 at shower compartments; reinforcement shall be provided for the future installation of grab bars and seats complying with those requirements.

EXCEPTIONS:

1. In a room containing only a lavatory and a water closet, reinforcement is not required provided the room does not contain the only lavatory or water closet on the accessible level of the unit.

2. At water closets reinforcement for the side wall vertical grab bar component required by Section 604.5 is not required.

3. At water closets where wall space will not permit a grab bar complying with Section 604.5.2, reinforcement for a rear wall grab bar 24 inches (610 mm) minimum in length centered on the water closet shall be provided.

4. At water closets where a side wall is not available for a 42-inch (1065 mm) grab bar complying with Section 604.5.1, reinforcement for a sidewall grab bar, 24 inches (610 mm) minimum in length, located 12 inches

(305 mm) maximum from the rear wall, shall be provided.

5. At water closets where a side wall is not available for a 42-inch (1065 mm) grab bar complying with Section 604.5.1 reinforcement for a swing-up grab bar complying with Section 1004.11.1.1 shall be permitted.

6. At water closets where a side wall is not available for a 42-inch (1065 mm) grab bar complying with Section 604.5.1 reinforcement for two swing-up grab bars complying with Section 1004.11.1.1 shall be permitted to be installed in lieu of reinforcement for rear wall and side wall grab bars.

7. In shower compartments larger than 36 inches (915 mm) in width and 36 inches (915 mm) in depth reinforcement for a shower seat is not required

1004.11.1.1 Swing–up Grab Bars. A clearance of 18 inches (455 mm) minimum from the centerline of the water closet to any side wall or obstruction shall be provided where reinforcement for swing–up grab bars is provided. When the approach to the water closet is from the side, the 18 inches (455 mm) minimum shall be on the side opposite the direction of approach. Reinforcement shall accommodate a swing–up grab bar centered 15$\frac{3}{4}$ inches (400 mm) from the centerline of the water closet and 28 inches (710 mm) minimum in length, measured from the wall to the end of the horizontal portion of the grab bar. Reinforcement shall accommodate a swing–up grab bar with a height in the down position of 33 inches (840 mm) minimum

and 36 inches (915 mm) maximum. Reinforcement shall be adequate to resist forces in accordance with Section 609.8.

EXCEPTION: Where a water closet is positioned with a wall to the rear and to one side, the centerline of the water closet shall be 16 inches (405 mm) minimum and 18 inches (455 mm) maximum from the sidewall.

1004.11.2 Clear Floor Space. Clear floor spaces required by Section 1004.11.3.1 (Option A) or 1004.11.3.2 (Option B) shall comply with Sections 1004.11.2 and 305.3.

1004.11.2.1 Doors. Doors shall not swing into the clear floor space or clearance for any fixture.

EXCEPTION: Where a clear floor space complying with Section 305.3, excluding knee and toe clearances under elements, is provided within the room beyond the arc of the door swing.

1004.11.2.2 Knee and Toe Clearance. Clear floor space at fixtures shall be permitted to include knee and toe clearances complying with Section 306.

1004.11.3 Toilet and Bathing Areas. Either all toilet and bathing areas provided shall comply with Section 1004.11.3.1 (Option A), or one toilet and bathing area shall comply with Section 1004.11.3.2 (Option B).

1004.11.3.1 Option A. Each fixture provided shall comply with Section 1004.11.3.1.

EXCEPTIONS:

1. Where multiple lavatories are provided in a single toilet and bathing area such that travel between fixtures does not require travel through other parts of the unit, not more than one lavatory is required to comply with Section 1004.11.3.1.

2. A lavatory and a water closet in a room containing only a lavatory and water closet, provided the room does not contain the only lavatory or water closet on the accessible level of the unit.

1004.11.3.1.1 Lavatory. A clear floor space complying with Section 305.3, positioned for a parallel approach, shall be provided at a lavatory. The clear floor space shall be centered on the lavatory.

EXCEPTION: A lavatory complying with Section 606 shall be permitted. Cabinetry shall be permitted under the lavatory provided the following criteria are met:

(a) The cabinetry can be removed without removal or replacement of the lavatory; and

(b) The floor finish extends under the cabinetry; and

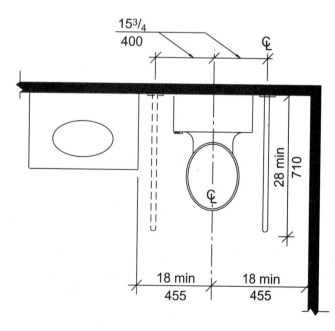

FIG. 1004.11.1.1
SWING-UP GRAB BAR FOR WATER CLOSET

(c) The walls behind and surrounding the cabinetry are finished.

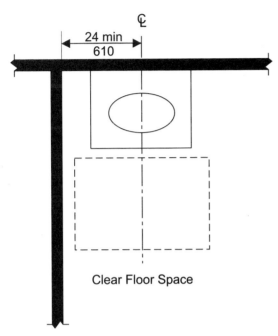

FIG. 1004.11.3.1.1
LAVATORY IN TYPE B UNITS—OPTION A BATHROOMS

1004.11.3.1.2 Water Closet. The water closet shall comply with Section 1004.11.3.1.2.

1004.11.3.1.2.1 Location. The centerline of the water closet shall be 16 inches (405 mm) minimum and 18 inches (455 mm) maximum from one side of the required clearance.

1004.11.3.1.2.2 Clearance. Clearance around the water closet shall comply with Sections 1004.11.3.1.2.2.1 through 1004.11.3.1.2.2.3.

EXCEPTION: Clearance complying with Sections 1003.11.2.4.2 through 1003.11.2.4.4.

1004.11.3.1.2.2.1 Clearance Width. Clearance around the water closet shall be 48 inches (1220 mm) minimum in width, measured perpendicular from the side of the clearance that is 16 inches (405 mm) minimum and 18 inches (455 mm) maximum from the water closet centerline.

1004.11.3.1.2.2.2 Clearance Depth. Clearance around the water closet shall be 56 inches (1420 mm) minimum in depth, measured perpendicular from the rear wall.

1004.11.3.1.2.2.3 Increased Clearance Depth at Forward Approach. Where a forward approach is provided, the clearance shall be 66 inches (1675 mm) minimum in depth, measured perpendicular from the rear wall.

1004.11.3.1.2.2.4 Clearance Overlap. A vanity or other obstruction 24 inches (610 mm) maximum in depth, measured perpendicular from the rear wall, shall be permitted to overlap the required clearance, provided the width of the remaining clearance at the water closet is 33 inches (840 mm) minimum.

1004.11.3.1.3 Bathing Fixtures. Where provided, a bathtub shall comply with Section 1004.11.3.1.3.1 or 1004.11.3.1.3.2 and a shower compartment shall comply with Section 1004.11.3.1.3.3.

1004.11.3.1.3.1 Parallel Approach Bathtubs. A clearance 60 inches (1525 mm) minimum in length and 30 inches (760 mm) minimum in width shall be provided in front of bathtubs with a parallel approach. Lavatories complying with Section 606 shall be permitted in the clearance. A lavatory complying with Section 1004.11.3.1.1 shall be permitted at one end of the bathtub if a clearance 48 inches (1220 mm) minimum in length and 30 inches (760 mm) minimum in width is provided in front of the bathtub.

1004.11.3.1.3.2 Forward Approach Bathtubs. A clearance 60 inches (1525 mm) minimum in length and 48 inches (1220 mm) minimum in width shall be provided in front of bathtubs with a forward approach. A water closet and a lavatory shall be permitted in the clearance at one end of the bathtub.

1004.11.3.1.3.3 Shower Compartment. If a shower compartment is the only bathing facility, the shower compartment shall have dimensions of 36 inches (915 mm) minimum in width and 36 inches (915 mm) minimum in depth. A clearance of 48 inches (1220 mm) minimum in length, measured perpendicular from the shower head wall, and 30 inches (760 mm) minimum in depth, measured from the face of the shower compartment, shall be provided. Reinforcing for a shower seat is not required in shower compartments larger than 36 inches (915 mm) in width and 36 inches (915 mm) in depth.

1004.11.3.2 Option B. One of each type of fixture provided shall comply with Section 1004.11.3.2. The accessible fixtures shall be in a single toilet/bathing area, such that travel between fixtures does not require travel through other parts of the unit.

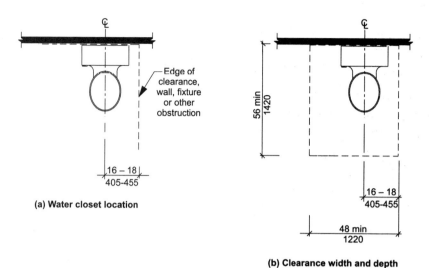

(a) Water closet location

(b) Clearance width and depth

(c) Increased clearance depth –
forward approach

(d) Clearance with lavatory overlap

FIG. 1004.11.3.1.2
CLEARANCE AT WATER CLOSETS IN TYPE B UNITS

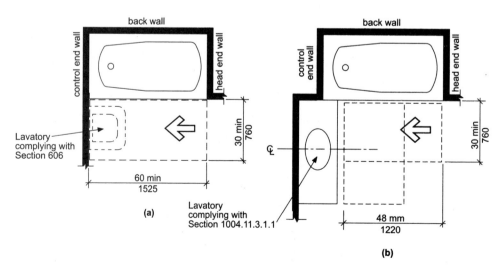

FIG. 1004.11.3.1.3.1
PARALLEL APPROACH BATHTUB IN TYPE B UNITS—OPTION A BATHROOMS

1004.11.3.2.1 Lavatory. Lavatories shall comply with Sections 1004.11.3.1.1 and 1004.11.3.2.1.1.

1004.11.3.2.1.1 Height. The front of the lavatory shall be 34 inches (865 mm) maximum above the floor, measured to the higher of the rim or counter surface.

1004.11.3.2.2 Water Closet. The water closet shall comply with Section 1004.11.3.1.2.

1004.11.3.2.3 Bathing Fixtures. The accessible bathing fixture shall be a bathtub complying with Section 1004.11.3.2.3.1 or a shower compartment complying with Section 1004.11.3.2.3.2.

1004.11.3.2.3.1 Bathtub. A clearance 48 inches (1220 mm) minimum in length measured perpendicular from the control end of the bathtub, and 30 inches (760 mm) minimum in width shall be provided in front of bathtubs.

1004.11.3.2.3.2 Shower Compartment. A shower compartment shall comply with Section 1004.11.3.1.3.3.

1004.12 Kitchens and kitchenettes. Kitchens and kitchenettes shall comply with Section 1004.12.

1004.12.1 Clearance. Clearance complying with Section 1004.12.1 shall be provided.

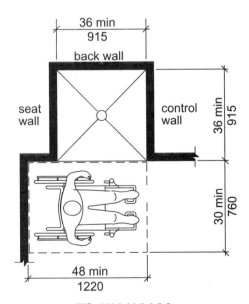

FIG. 1004.11.3.1.3.2
FORWARD APPROACH BATHTUB IN
TYPE B UNITS—OPTION A BATHROOMS

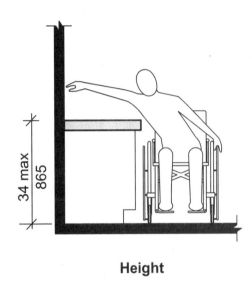

Height

FIG. 1004.11.3.2.1
LAVATORY IN TYPE B UNITS—OPTION B BATHROOMS

FIG. 1004.11.3.1.3.3
TRANSFER-TYPE SHOWER
COMPARTMENT IN TYPE B UNITS

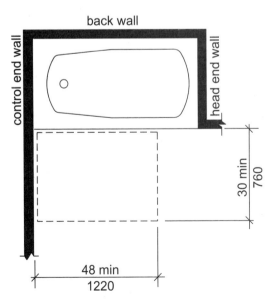

FIG. 1004.11.3.2.3.1
BATHROOM CLEARANCE IN
TYPE B UNITS—OPTION B BATHROOMS

1004.12.1.1 Minimum Clearance. Clearance between all opposing base cabinets, counter tops, appliances, or walls within kitchen work areas shall be 40 inches (1015mm) minimum.

1004.12.1.2 U-Shaped Kitchens. In kitchens with counters, appliances, or cabinets on three contiguous sides, clearance between all opposing base cabinets, countertops, appliances, or walls within kitchen work areas shall be 60 inches (1525 mm) minimum.

1004.12.2 Clear Floor Space. Clear floor space at appliances shall comply with Sections 1004.12.2 and 305.3.

1004.12.2.1 Sink. A clear floor space, positioned for a parallel approach to the sink, shall be provided. The clear floor space shall be centered on the sink bowl.

> **EXCEPTION:** A sink with a forward approach complying with Section 1003.12.4.1.

1004.12.2.2 Dishwasher. A clear floor space, positioned for a parallel or forward approach to the dishwasher, shall be provided. The dishwasher door in the open position shall not obstruct the clear floor space for the dishwasher.

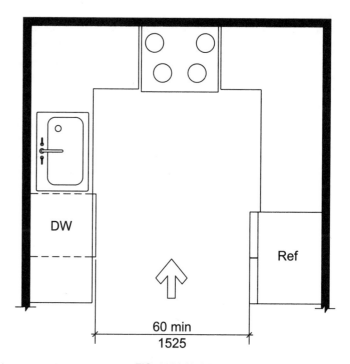

FIG. 1004.12.1.2
U-SHAPED KITCHEN CLEARANCE IN TYPE B UNITS

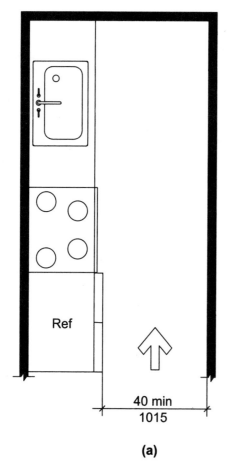

(a)

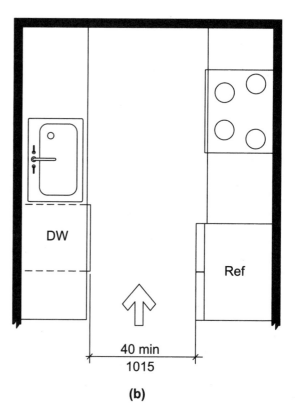

(b)

FIG. 1004.12.1.1
MINIMUM KITCHEN CLEARANCE IN TYPE B UNITS

1004.12.2.3 Cooktop. Cooktops shall comply with Section 1004.12.2.3.

1004.12.2.3.1 Approach. A clear floor space, positioned for a parallel or forward approach to the cooktop, shall be provided.

1004.12.2.3.2 Forward approach. Where the clear floor space is positioned for a forward approach, knee and toe clearance complying with Section 306 shall be provided. The underside of the cooktop shall be insulated or otherwise configured to prevent burns, abrasions, or electrical shock.

1004.12.2.3.3 Parallel approach. Where the clear floor space is positioned for a parallel approach, the clear floor space shall be centered on the appliance.

1004.12.2.4 Oven. A clear floor space, positioned for a parallel or forward approach adjacent to the oven shall be provided. The oven door in the open position shall not obstruct the clear floor space for the oven.

1004.12.2.5 Refrigerator/Freezer. A clear floor space, positioned for a parallel approach to the refrigerator/freezer, shall be provided. The center-line of the clear floor space shall be offset 24 inches (610 mm) maximum from the centerline of the appliance.

1004.12.2.6 Trash Compactor. A clear floor space, positioned for a parallel or forward approach to the trash compactor, shall be provided.

1005 Type C (Visitable) Units

1005.1 General. Type C (Visitable) dwelling units shall comply with Section 1005.

1005.2 Unit Entrance. At least one unit entrance shall be on a circulation path complying with Section 1005.5 from a public street or sidewalk, a dwelling unit driveway, or a garage.

1005.3 Connected Spaces. A circulation path complying with Section 1005.5 shall connect the unit entrance complying with Section 1005.2 and with the spaces specified in Section 1005.4.

1005.4 Interior Spaces. The entrance level shall include a toilet room or bathroom complying with Section 1005.6 and one habitable space with an area 70 square feet (6.5 m²) minimum. Where a food preparation area is provided on the entrance level, it shall comply with Section 1005.7.

Exception: A toilet room or bathroom shall not be required on an entrance level with less than 120 square feet (11.1 m²) of habitable space.

1005.5 Circulation Path. Circulation paths shall comply with Section 1005.5.

1005.5.1 Components. The circulation path shall consist of one or more of the following elements:

walking surfaces with a slope not steeper that 1:20, doors and doorways, ramps, elevators complying with Sections 407 through 409, and wheelchair (platform) lifts complying with Section 410.

1005.5.2 Walking Surfaces. Walking surfaces with slopes not steeper than 1:20 shall comply with Section 303.

1005.5.2.1 Clear Width. The clear width of the circulation path shall comply with Section 403.5.

1005.5.3 Doors and Doorways. Doors and doorways shall comply with Section 1005.5.3

1005.5.3.1 Clear Width. Doorways shall have a clear opening of 31³/₄ inches (805 mm) minimum. Clear opening of swinging doors shall be measured between the face of the door and stop, with the door open 90 degrees.

1005.5.3.2 Thresholds. Thresholds shall comply with Section 303.

Exception: Thresholds at exterior sliding doors shall be permitted to be ³/₄ inch (19 mm) maximum in height, provided they are beveled with a slope not steeper than 1:2.

1005.5.4 Ramps. Ramps shall comply with Section 405.

Exception: Handrails, intermediate landings and edge protection are not required where the sides of ramp runs have a vertical drop off of ¹/₂ inch (13 mm) maximum within 10 inches (255 mm) horizontally of the ramp run.

1005.5.4.1 Clear Width. The clear width of the circulation path shall comply with Section 403.5.

1005.6 Toilet Room or Bathroom. At a minimum, the toilet room or bathroom required by Section 1005.4 shall include a lavatory and a water closet. Reinforcement shall be provided for the future installation of grab bars at water closets. Clearances at the water closet shall comply with Section 1004.11.3.1.2.

1005.7 Food Preparation Area. At a minimum, the food preparation area shall include a sink, a cooking appliance, and a refrigerator. Clearances between all opposing base cabinets, counter tops, appliances or walls within the food preparation area shall be 40 inches (1015 mm) minimum in width.

Exception: Spaces that do not provide a cooktop or conventional range shall be permitted to provide a clearance of 36 inches (915 mm) minimum in width.

1005.8 Lighting Controls and Receptacle Outlets. Receptacle outlets and operable parts of lighting controls shall be located 15 inches (380 mm) minimum and 48 inches (1220 mm) maximum above the floor.

Exception: The following shall not be required to comply with Section 1005.8.

1. Receptacle outlets serving a dedicated use.

2. Controls mounted on ceiling fans and ceiling lights.

3. Floor receptacle outlets.

4. Lighting controls and receptacle outlets over countertops.

1006 Units with Accessible Communication Features

1006.1 General. Units required to have accessible communication features shall comply with Section 1006.

1006.2 Unit Smoke Detection. Where provided, unit smoke detection shall include audible notification complying with NFPA 72 listed in Section 105.2.2.

1006.3 Building Fire Alarm System. Where a building fire alarm system is provided, the system wiring shall be extended to a point within the unit in the vicinity of the unit smoke detection system.

1006.4 Visible Notification Appliances. Visible notification appliances, where provided within the unit as part of the unit smoke detection system or the building fire alarm system, shall comply with Section 1006.4.

1006.4.1 Appliances. Visible notification appliances shall comply with Section 702.

1006.4.2 Activation. All visible notification appliances provided within the unit for smoke detection notification shall be activated upon smoke detection. All visible notification appliances provided within the unit for building fire alarm notification shall be activated upon activation of the building fire alarm in the portion of the building containing the unit.

1006.4.3 Interconnection. The same visible notification appliances shall be permitted to provide notification of unit smoke detection and building fire alarm activation.

1006.4.4 Prohibited Use. Visible notification appliances used to indicate unit smoke detection or building fire alarm activation shall not be used for any other purpose within the unit.

1006.5 Unit Primary Entrance. Communication features shall be provided at the unit primary entrance complying with Section 1006.5.

1006.5.1 Notification. A hard-wired electric doorbell shall be provided. A button or switch shall be provided on the public side of the unit primary entrance. Activation of the button or switch shall initiate an audible tone within the unit.

1006.5.2 Identification. A means for visually identifying a visitor without opening the unit entry door shall be provided. Peepholes, where used, shall provide a minimum 180-degree range of view.

1006.6 Site, Building, or Floor Entrance. Where a system permitting voice communication between a visitor and the occupant of the unit is provided at a location other than the unit entry door, the system shall comply with Section 1006.6.

1006.6.1 Public or Common-Use Interface. The public or common-use system interface shall include the capability of supporting voice and TTY communication with the unit interface.

1006.6.2 Unit Interface. The unit system interface shall include a telephone jack capable of supporting voice and TTY communication with the public or common-use system interface.

1006.7 Closed-Circuit Communication Systems. Where a closed-circuit communication system is provided, the public or common-use system interface shall comply with Section 1006.6.1, and the unit system interface in units required to have accessible communication features shall comply with Section 1006.6.2.

Chapter 11. Recreational Facilities

1101 General

1101.1 Scope. Recreational facilities required to be accessible by the scoping provisions adopted by the administrative authority shall comply with the applicable provisions of Chapter 11.

1101.2 Special Provisions.

1101.2 .1 General Exceptions. The following shall not be required to be accessible or to be on an accessible route:

1. Raised structures used solely for refereeing, judging, or scoring a sport.

2. Water Slides.

3. Animal containment areas that are not for public use.

4. Raised boxing or wrestling rings.

5. Raised diving boards and diving platforms.

6. Bowling lanes that are not required to provide wheelchair spaces.

7. Mobile or portable amusement rides

8. Amusement rides that are controlled or operated by the rider.

9. Amusement rides designed primarily for children, where children are assisted on and off the ride by an adult.

10. Amusement rides that do not provide amusement ride seats.

1101.2.2 Area of Sport Activity. Areas of sport activity shall be served by an accessible route and shall not be required to be accessible except as provided in Chapter 11.

1101.2.3 Recreational Boating Facilities. Operable parts of cleats and other boat securement devices shall not be required to comply with Section 308.

1101.2.4 Exercise Machines and Equipment. Exercise machines and exercise equipment shall not be required to comply with Section 309.

1101.3 Protruding Objects. Protruding objects on circulation paths shall comply with Section 307.

EXCEPTIONS:

1. Within areas of sport activity, protruding objects on circulation paths shall not be required to comply with Section 307.

2. Within play areas, protruding objects on circulation paths shall not be required to comply with Section 307 provided that ground level accessible routes provide vertical clearance complying with Section 1108.2.

1102 Amusement Rides.

1102.1 General. Accessible amusement rides shall comply with Section 1102.

1102.2 Accessible Routes. Accessible routes serving amusement rides shall comply with Chapter 4.

EXCEPTIONS:

1. In load or unload areas and on amusement rides, where complying with Section 405.2 is not structurally or operationally feasible, ramp slope shall be permitted to be 1:8 maximum.

2. In load or unload areas and on amusement rides, handrails provided along walking surfaces complying with Section 403 and required on ramps complying with Section 405 shall not be required to comply with Section 505 where complying is not structurally or operationally feasible.

1102.3 Load and Unload Areas. A turning space complying with Sections 304.2 and 304.3 shall be provided in load and unload areas.

1102.4 Wheelchair Spaces in Amusement Rides. Wheelchair spaces in amusement rides shall comply with Section 1102.4.

1102.4.1 Floor Surface. The floor surface of wheelchair spaces shall be stable and firm.

1102.4.2 Slope. The floor surface of wheelchair spaces shall have a slope not steeper than 1:48 when in the load and unload position.

1102.4.3 Gaps. Floors of amusement rides with wheelchair spaces and floors of load and unload areas shall be coordinated so that, when amusement rides are at rest in the load and unload position, the vertical difference between the floors shall be within plus or minus $\frac{5}{8}$ inch (16 mm) and the horizontal gap shall be 3 inches (75 mm) maximum under normal passenger load conditions.

EXCEPTION: Where complying is not operationally or structurally feasible, ramps, bridge plates, or similar devices complying with the applicable requirements of 36 CFR 1192.83(c), listed in Section 105.2.11, shall be provided.

1102.4.4 Clearances. Clearances for wheelchair spaces shall comply with Section 1102.4.4.

EXCEPTIONS:

1. Where provided, securement devices shall be permitted to overlap required clearances.

2. Wheelchair spaces shall be permitted to be mechanically or manually repositioned.

3. Wheelchair spaces shall not be required to comply with Section 307.4.

1102.4.4.1 Width and Length. Wheelchair spaces shall provide a clear width of 30 inches (760 mm) minimum and a clear length of 48 inches (1220 mm) minimum measured to 9 inches (230 mm) minimum above the floor.

1102.4.4.2 Side Entry. Where wheelchair spaces are entered only from the side, amusement rides shall be designed to permit sufficient maneuvering clearance for individuals using a wheelchair or mobility aid to enter and exit the ride.

1102.4.4.3 Permitted Protrusions in Wheelchair Spaces. Objects are permitted to protrude a distance of 6 inches (150 mm) maximum along the front of the wheelchair space, where located 9 inches (230 mm) minimum and 27 inches (685 mm) maximum above the floor of the wheelchair space. Objects are permitted to protrude a distance of 25 inches (635 mm) maximum along the front of the wheelchair space, where located more than 27 inches (685 mm) above the floor of the wheelchair space.

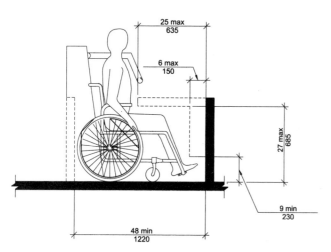

**FIGURE 1102.4.4.3
PROTRUSIONS IN WHEELCHAIR
SPACES IN AMUSEMENT RIDES**

1102.4.5 Ride Entry. Openings providing entry to wheelchair spaces on amusement rides shall provide a clear width of 32 inches (815 mm) minimum.

1102.4.6 Approach. One side of the wheelchair space shall adjoin an accessible route when in the load and unload position.

1102.4.7 Companion Seats. Where the interior width of the amusement ride is greater than 53 inches (1345 mm), seating is provided for more than one rider, and the wheelchair is not required to be centered within the amusement ride, a companion seat shall be provided for each wheelchair space.

1102.4.7.1 Shoulder-to-Shoulder Seating. Where an amusement ride provides shoulder-to-shoulder seating, companion seats shall be shoul-der-to-shoulder with the adjacent wheelchair space.

> **EXCEPTION:** Where shoulder-to-shoulder companion seating is not operationally or structurally feasible, complying with this requirement shall be required to the maximum extent practicable.

1102.5 Amusement Ride Seats Designed for Transfer. Amusement ride seats designed for transfer shall comply with Section 1102.5 when positioned for loading and unloading.

1102.5.1 Clear Floor Space. A clear floor space complying with Section 305 shall be provided in the load and unload area adjacent to the amusement ride seats designed for transfer.

1102.5.2 Transfer Height. The height of amusement ride seats designed for transfer shall be 14 inches (355 mm) minimum and 24 inches (610 mm) maximum measured from the surface of the load and unload area.

1102.5.3 Transfer Entry. Where openings are provided for transfer to amusement ride seats, the openings shall provide clearance for transfer from a wheelchair or mobility aid to the amusement ride seat.

1102.5.4 Wheelchair Storage Space. Wheelchair storage spaces complying with Section 305 shall be provided in or adjacent to unload areas for each required amusement ride seat designed for transfer and shall not overlap any required means of egress or accessible route.

1102.6 Transfer Devices for Use with Amusement Rides. Transfer devices for use with amusement rides shall comply with Section 1102.6 when positioned for loading and unloading.

1102.6.1 Clear Floor Space. A clear floor space complying with Section 305 shall be provided in the load and unload area adjacent to the transfer device.

1102.6.2 Transfer Height. The height of transfer device seats shall be 14 inches (355 mm) minimum and 24 inches (610 mm) maximum measured from the load and unload surface.

1102.6.3 Wheelchair Storage Space. Wheelchair storage spaces complying with Section 305 shall be provided in or adjacent to unload areas for each required transfer device and shall not overlap any required means of egress or accessible route.

1103 Recreational Boating Facilities

1103.1 General. Accessible recreational boating facilities shall comply with Section 1103.

1103.2 Accessible Routes. Accessible routes serving recreational boating facilities, including gangways and floating piers, shall comply with Chapter 4 except as modified by the exceptions in Section 1103.2.

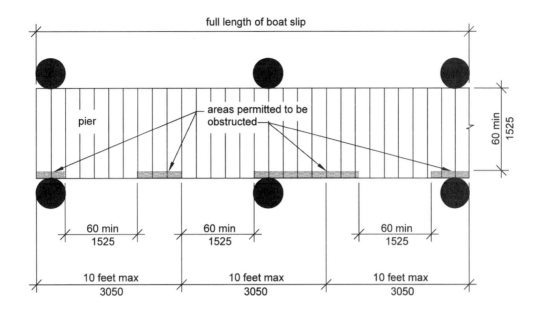

FIGURE 1103.3.1(A)
BOAT SLIP CLEARANCE

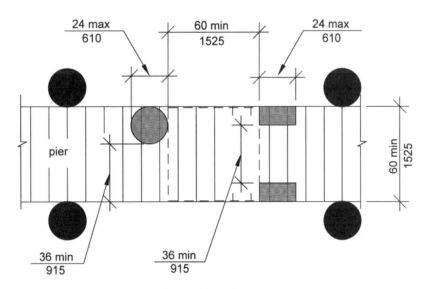

FIGURE 1103.3.1(B)
(EXCEPTION 1) CLEAR PIER SPACE REDUCTION AT BOAT SLIPS

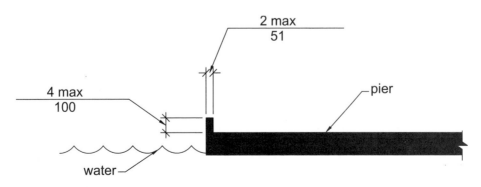

FIGURE 1103.3.1(C)
(EXCEPTION 2) EDGE PROTECTION AT BOAT SLIPS

1103.2.1 Boat Slips. An accessible route shall serve boat slips.

EXCEPTIONS:

1. Where an existing gangway or series of gangways is replaced or altered, an increase in the length of the gangway shall not be required to comply with Section 1103.2.

2. Gangways shall not be required to comply with the maximum rise specified in Section 405.6.

3. Where the total length of a gangway or series of gangways serving as part of a required accessible route is 80 feet (24 m) minimum, gangways shall not be required to comply with Section 405.2.

4. Where facilities contain fewer than 25 boat slips and the total length of the gangway or series of gangways serving as part of a required accessible route is 30 feet (9145 mm) minimum, gangways shall not be required to comply with Section 405.2.

5. Where gangways connect to transition plates, landings specified by Section 405.7 shall not be required.

6. Where gangways and transition plates connect and are required to have handrails, handrail extensions shall not be required. Where handrail extensions are provided on gangways or transition plates, the handrail extensions shall not be required to be parallel with the floor.

7. The cross slope specified in Sections 403.3 and 405.3 for gangways, transition plates, and floating piers that are part of accessible routes shall be measured in the static position.

8. Changes in level complying with Sections 303.3 and 303.4 shall be permitted on the surfaces of gangways and boat launch ramps.

9. Cleats and other boat securement devices shall not be required to comply with Section 309.3.

1103.2.2 Boarding Piers at Boat Launch Ramps. An accessible route shall serve boarding piers.

EXCEPTIONS:

1. Accessible routes serving floating boarding piers shall be permitted to use Exceptions 1, 2, 5, 6, 7 and 8 in Section1103.2.1.

2. Where the total length of the gangway or series of gangways serving as part of a required accessible route is 30 feet (9145 mm) minimum, gangways shall not be required to comply with Section 405.2.

3. Where the accessible route serving a floating boarding pier or skid pier is located within a boat launch ramp, the portion of the accessible route located within the boat launch ramp shall not be required to comply with Section 405.

1103.3 Clearances. Clearances at boat slips and on boarding piers at boat launch ramps shall comply with Section 1103.3.

1103.3.1 Boat Slip Clearance. Boat slips shall provide clear pier space 60 inches (1525 mm) minimum in width that extend the full length of the boat slips. Each 10 feet (3050 mm) of linear pier edge serving boat slips shall contain at least one continuous clear opening 60 inches (1525 mm) minimum in width.

EXCEPTIONS:

1. Clear pier space shall be permitted to be 36 inches (915 mm) minimum in width and 24 inches (610 mm) maximum in length, provided that multiple 36-inch (915 mm) wide segments are separated by segments that are 60 inches (1525 mm) minimum in width and 60 inches (1525 mm) minimum in length.

2. Edge protection shall be permitted at the continuous clear openings, provided the edge protection is 4 inches (100 mm) maximum in height and 2 inches (51 mm) maximum in width.

3. In existing piers, clear pier space shall be permitted to be located perpendicular to the boat slip and shall extend the width of the boat slip, where the facility has at least one boat slip complying with Section 1103.3, and further compliance with Section 1103.3 would result in a reduction in the number of boat slips available or result in a reduction of the widths of existing slips.

1103.3.2 Boarding Pier Clearances. Boarding piers at boat launch ramps shall provide clear pier space 60 inches (1525 mm) minimum in width and shall extend the full length of the boarding pier. Every 10 feet (3050 mm) of linear pier edge shall contain at least one continuous clear opening 60 inches (1525 mm) minimum in width.

EXCEPTIONS:

1. The clear pier space shall be permitted to be 36 inches (915 mm) minimum in width and 24 inches (610 mm) maximum in length provided that multiple 36-inch (915 mm) wide segments are separated by segments that are 60 inches (1525 mm) minimum in width and 60 inches (1525 mm) minimum in length.

2. Edge protection shall be permitted at the continuous clear openings provided the edge protection is 4 inches (100 mm) maxi-

mum in height and 2 inches (51 mm) maximum in width.

1104 Exercise Machines and Equipment

1104.1 Clear Floor Space. Accessible exercise machines and equipment shall have a clear floor space complying with Section 305 positioned for transfer or for use by an individual seated in a wheelchair. Clear floor spaces required at exercise machines and equipment shall be permitted to overlap.

1105 Fishing Piers and Platforms

1105.1 Accessible Routes. Accessible routes serving fishing piers and platforms, including gangways and floating piers, shall comply with Chapter 4.

EXCEPTIONS:

1. Accessible routes serving floating fishing piers and platforms shall be permitted to use Exceptions 1, 2, 5, 6, 7 and 8 in Section 1103.2.1.

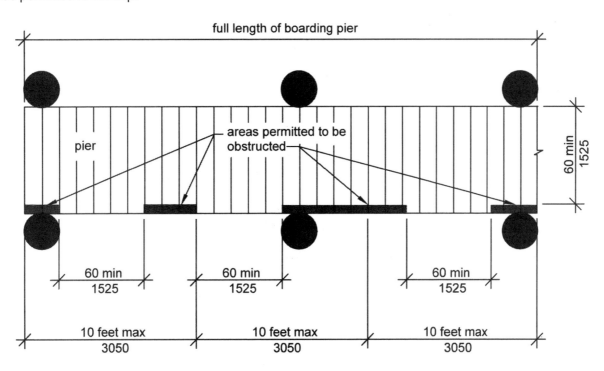

FIGURE 1103.3.2(A)
BOARDING PIER CLEARANCE

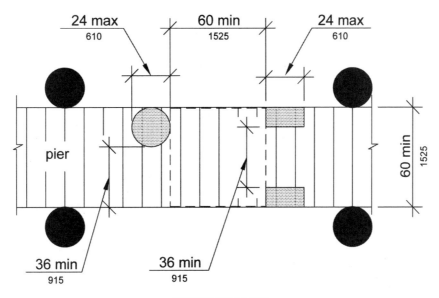

FIGURE 1103.3.2(B)
(EXCEPTION 1) CLEAR PIER SPACE REDUCTION AT BOARDING PIERS

2. Where the total length of the gangway or series of gangways serving as part of a required accessible route is 30 feet (9145 mm) minimum, gangways shall not be required to comply with Section 405.2.

1105.2 Railings. Where provided, railings, guards, or handrails shall comply with Section 1105.2.

1105.2.1 Height. A minimum of 25 percent of the railings, guards, or handrails shall be 34 inches (865 mm) maximum above the ground or deck surface.

EXCEPTION: Where a guard complying with the applicable building code is provided, the guard shall not be required to comply with Section 1105.2.1.

1105.2.1.1 Dispersion. Railings, guards, or handrails required to comply with Section 1105.2.1 shall be dispersed throughout the fishing pier or platform.

1105.3 Edge Protection. Where railings, guards, or handrails complying with Section 1105.2 are provided, edge protection complying with Section 1105.3.1 or 1105.3.2 shall be provided.

1105.3.1 Curb or Barrier. Curbs or barriers shall extend 2 inches (51 mm) minimum in height above the surface of the fishing pier or platform.

1105.3.2 Extended Ground or Deck Surface. The ground or deck surface shall extend 12 inches (305 mm) minimum beyond the inside face of the railing. Toe clearance shall be provided and shall be 30 inches (760 mm) minimum in width and 9 inches (230 mm) minimum in height above the ground or deck surface beyond the railing.

1105.4 Clear Floor Space. At each location where there are railings, guards, or handrails complying with Section 1105.2.1, a clear floor space complying with Section 305 shall be provided. Where there are no railings, guards, or handrails, at least one clear floor space complying with Section 305 shall be provided on the fishing pier or platform.

1105.5 Turning Space. At least one turning space complying with Section 304.3 shall be provided on fishing piers and platforms.

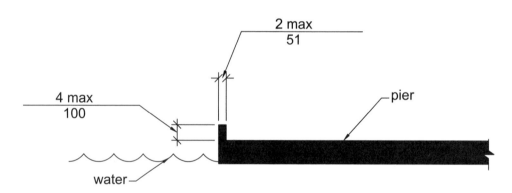

FIGURE 1103.3.2(C)
(EXCEPTION 2) EDGE PROTECTION AT BOARDING PIERS

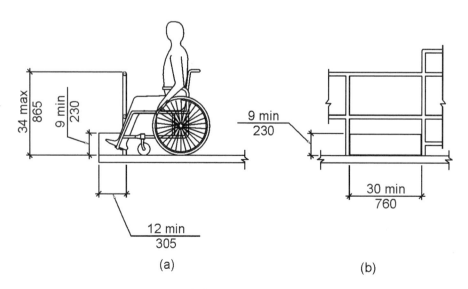

(a)   (b)

FIGURE 1105.3.2
EXTENDED GROUND OR DECK SURFACE AT FISHING PIERS AND PLATFORMS

1106 Golf Facilities

1106.1 General. Golf facilities shall comply with Section 1106.

1106.2 Accessible Routes. Accessible routes serving teeing grounds, practice teeing grounds, putting greens, practice putting greens, teeing stations at driving ranges, course weather shelters, golf car rental areas, bag drop areas, and course toilet rooms shall comply with Chapter 4 and shall be 48 inches (1220 mm) minimum in width. Where handrails are provided, accessible routes shall be 60 inches (1525 mm) minimum in width.

> **EXCEPTION:** Handrails shall not be required on golf courses. Where handrails are provided on golf courses, the handrails shall not be required to comply with Section 505.

1106.3 Golf Car Passages. Golf car passages shall comply with Section 1106.3.

1106.3.1 Clear Width. The clear width of golf car passages shall be 48 inches (1220 mm) minimum.

1106.3.2 Barriers. Where curbs or other constructed barriers prevent golf cars from entering a fairway, openings 60 inches (1525 mm) minimum in width shall be provided at intervals not to exceed 75 yards (69 m).

1106.4 Weather Shelters. A clear floor space 60 inches (1525 mm) minimum by 96 inches (2440 mm) minimum shall be provided within weather shelters.

1107 Miniature Golf Facilities

1107.1 General. Miniature golf facilities shall comply with Section 1107.

1107.2 Accessible Routes. Accessible routes serving holes on miniature golf courses shall comply with Chapter 4.

> **EXCEPTION:** Accessible routes located on playing surfaces of miniature golf holes shall be permitted to comply with the following:
>
> 1. Playing surfaces shall not be required to comply with Section 302.2.
>
> 2. Where accessible routes intersect playing surfaces of holes, a curb that is 1 inch (25 mm) maximum in height and 32 inches (815 mm) minimum in width shall be permitted.
>
> 3. A slope of 1:4 maximum shall be permitted for a rise of 4 inches (100 mm) maximum.
>
> 4. Ramp landing slopes specified by Section 405.7.1 shall be permitted to be 1:20 maximum.
>
> 5. Ramp landing length specified by Section 405.7.3 shall be permitted to be 48 inches (1220 mm) minimum.
>
> 6. Ramp landing size at a change in direction specified by Section 405.7.4 shall be permitted

to be 48 inches (1220 mm) minimum by 60 inches (1525 mm) minimum.

> 7. Handrails shall not be required on holes. Where handrails are provided on holes, the handrails shall not be required to comply with Section 505.

1107.3 Miniature Golf Holes. Miniature golf holes shall comply with Section 1107.3.

1107.3.1 Start of Play. A clear floor space 48 inches (1220 mm) minimum by 60 inches (1525 mm) minimum with slopes not steeper than 1:48 shall be provided at the start of play.

1107.3.2 Golf Club Reach Range Area. All areas within holes where golf balls rest shall be within 36 inches (915 mm) maximum of a clear floor space 36 inches (915 mm) minimum in width and 48 inches (1220 mm) minimum in length having a running slope not steeper than 1:20. The clear floor space shall be served by an accessible route.

1108 Play Areas

1108.1 Scope. Play areas shall comply with 1108.

1108.2 Accessible Routes for Play Areas. Play areas shall provide accessible routes in accordance with Section 1108.2. Accessible routes serving play areas shall comply with Chapter 4 except as modified by Section 1108.4.

1108.2.1 Ground Level and Elevated Play Components. At least one accessible route shall be provided within the play area. The accessible route shall connect ground level play components required to comply with Section 1108.3.2.1 and elevated play components required to comply with Section 1108.3.2.2, including entry and exit points of the play components.

1108.2.2 Soft Contained Play Structures. Where three or fewer entry points are provided for soft contained play structures, at least one entry point shall be on an accessible route. Where four or more entry points are provided for soft contained play structures, at least two entry points shall be on an accessible route.

1108.3 Age Groups. Play areas for children ages 2 and over shall comply with Section 1108.3. Where separate play areas are provided within a site for specific age groups, each play area shall comply with Section 1108.3.

> **EXCEPTIONS:**
>
> 1. Play areas located in family child care facilities where the proprietor actually resides shall not be required to comply with Section 1108.3.
>
> 2. In existing play areas, where play components are relocated for the purposes of creating safe use zones and the ground surface is not altered or extended for more than one use

zone, the play area shall not be required to comply with Section 1108.3.

3. Amusement attractions shall not be required to comply with Section 1108.3.

4. Where play components are altered and the ground surface is not altered, the ground surface shall not be required to comply with Section 1108.4.1.6 unless required by the authority having jurisdiction.

1108.3.1 Additions. Where play areas are designed and constructed in phases, the requirements of Section 1108.3 shall apply to each successive addition so that when the addition is completed, the entire play area complies with all the applicable requirements of Section 1108.3.

1108.3.2 Play Components. Where provided, play components shall comply with Section 1108.3.2.

1108.3.2.1 Ground Level Play Components. Ground level play components shall be provided in the number and types required by Section 1108.3.2.1. Ground level play components that are provided to comply with Section 1108.3.2.1.1 shall be permitted to satisfy the additional number required by Section 1108.3.2.1.2 if the minimum required types of play components are satisfied. Where two or more required ground level play components are provided, they shall be dispersed throughout the play area and integrated with other play components.

1108.3.2.1.1 Minimum Number and Types. Where ground level play components are provided, at least one of each type shall be on an accessible route and shall comply with Section 1108.4.3.

1108.3.2.1.2 Additional Number and Types. Where elevated play components are provided,

ground level play components shall be provided in accordance with Table 1108.3.2.1.2 and shall comply with Section 1108.4.3.

EXCEPTION: If at least 50 percent of the elevated play components are connected by a ramp and at least 3 of the elevated play components connected by the ramp are different types of play components, the play area shall not be required to comply with Section 1108.3.2.1.2.

1108.3.2.2 Elevated Play Components. Where elevated play components are provided, at least 50 percent shall be on an accessible route and shall comply with Section 1108.4.3.

1108.4 Accessible Routes Within Play areas. Play areas shall comply with Section 1108.4.

1108.4.1 Accessible Routes. Accessible routes serving play areas shall comply with Chapter 4 and Section 1108.4.1 and shall be permitted to use the exceptions in Sections 1108.4.1.1 through 1108.4.1.3. Where accessible routes serve ground level play components, the vertical clearance shall be 80 inches (2030 mm) minimum in height.

1108.4.1.1 Ground Level and Elevated Play Components. Accessible routes serving ground level play components and elevated play components shall be permitted to use the exceptions in Section 1108.4.1.1.

EXCEPTIONS:

1. Transfer systems complying with Section 1108.4.2 shall be permitted to connect elevated play components except where 20 or more elevated play components are provided no more than 25 percent of the elevated play components shall be

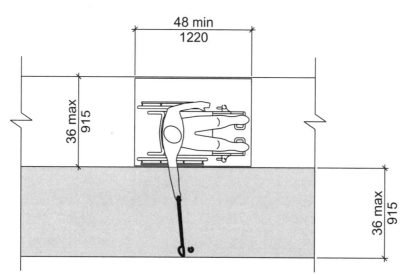

Note: Running Slope of Clear Floor or Ground Space Not Steeper Than 1:20

FIGURE 1107.3.2
GOLF CLUB REACH RANGE AREA

permitted to be connected by transfer systems.

2. Where transfer systems are provided, an elevated play component shall be permitted to connect to another elevated play component as part of an accessible route.

1108.4.1.2 Soft Contained Play Structures. Accessible routes serving soft contained play structures shall be permitted to use the exception in Section 1108.4.1.2.

EXCEPTION: Transfer systems complying with Section 1108.4.2 shall be permitted to be used as part of an accessible route.

1108.4.1.3 Water Play Components. Accessible routes serving water play components shall be permitted to use the exceptions in Section 1108.4.1.3.

EXCEPTIONS:

1. Where the surface of the accessible route, clear floor spaces, or turning spaces serving water play components is submerged, complying with Sections 302, 403.3, 405.2, 405.3, and 1108.4.1.6 shall not be required.

2. Transfer systems complying with Section 1108.4.2 shall be permitted to connect elevated play components in water.

1108.4.1.4 Clear Width. Accessible routes connecting play components shall provide a clear width complying with Section 1108.4.1.4.

1108.4.1.4.1 Ground Level. At ground level, the clear width of accessible routes shall be 60 inches (1525 mm) minimum.

EXCEPTIONS:

1. In play areas less than 1000 square feet (93 m²), the clear width of accessible routes shall be permitted to be 44 inches (1120 mm) minimum, if at least one turning space complying with Section 304.3 is provided where the restricted accessible route exceeds 30 feet (9145 mm) in length.

2. The clear width of accessible routes shall be permitted to be 36 inches (915 mm) minimum for a distance of 60 inches (1525 mm) maximum provided that multiple reduced width segments are separated by segments that are 60 inches (1525 mm) minimum in width and 60 inches (1525 mm) minimum in length.

1108.4.1.4.2 Elevated. The clear width of accessible routes connecting elevated play components shall be 36 inches (915 mm) minimum.

EXCEPTIONS:

1. The clear width of accessible routes connecting elevated play components shall be permitted to be reduced to 32 inches (815 mm) minimum for a distance of 24 inches (610 mm) maximum provided that reduced width segments are separated by segments that are 48 inches (1220 mm) minimum in length and 36 inches (915 mm) minimum in width.

2. The clear width of transfer systems connecting elevated play components shall be permitted to be 24 inches (610 mm) minimum.

1108.4.1.5 Ramps. Within play areas, ramps connecting ground level play components and ramps connecting elevated play components shall comply with Section 1108.4.1.5.

1108.4.1.5.1 Ground Level. Ramp runs connecting ground level play components shall have a running slope not steeper than 1:16.

TABLE 1108.3.2.1.2 NUMBER AND TYPES OF GROUND LEVEL PLAY COMPONENTS REQUIRED TO BE ON ACCESSIBLE ROUTES

Number of Elevated Play Components Provided	Minimum Number of Ground Level Play Components Required to be on an Accessible Route	Minimum Number of Different Types of Ground Level Play Components Required to be on an Accessible Route
1	Not applicable	Not applicable
2 to 4	1	1
5 to 7	2	2
8 to 10	3	3
11 to 13	4	3
14 to 16	5	3
17 to 19	6	3
22 to 22	7	4
23 to 25	8	4
26 and over	8, plus 1 for each additional 3, or fraction thereof, over 25	5

1108.4.1.5.2 Elevated. The rise for any ramp run connecting elevated play components shall be 12 inches (305 mm) maximum.

1108.4.1.5.3 Handrails. Where required on ramps serving play components, the handrails shall comply with Section 505 except as modified by Section 1108.4.1.5.3.

EXCEPTIONS:

1. Handrails shall not be required on ramps located within ground level use zones.

2. Handrail extensions shall not be required.

1108.4.1.5.3.1 Handrail Gripping Surfaces. Handrail gripping surfaces with a circular cross section shall have an outside diameter of 0.95 inch (24 mm) minimum and 1.55 inches (39 mm) maximum. Where the shape of the gripping surface is noncircular, the handrail shall provide an equivalent gripping surface.

1108.4.1.5.3.2 Handrail Height. The top of handrail gripping surfaces shall be 20 inches (510 mm) minimum and 28 inches (710 mm) maximum above the ramp surface.

1108.4.1.6 Ground Surfaces. Ground surfaces on accessible routes, clear floor spaces, and turning spaces shall comply with Section 1108.4.1.6.

1108.4.1.6.1 Surface Condition. Ground surfaces shall be stable, firm and slip resistant. Ground surfaces shall be inspected and maintained regularly and frequently to ensure continued compliance with this requirement.

1108.4.1.6.2 Use Zones. Ground surfaces located within use zones shall comply with ASTM F 1292 listed in Sections 105.2.8 or 105.2.9.

1108.4.2 Transfer Systems. Where transfer systems are provided to connect to elevated play components, the transfer systems shall comply with Section 1108.4.2.

1108.4.2.1 Transfer Platforms. Transfer platforms shall be provided where transfer is intended from wheelchairs or other mobility aids. Transfer platforms shall comply with Section 1108.4.2.1.

1108.4.2.1.1 Size. Transfer platforms shall have level surfaces 14 inches (355 mm) minimum in depth and 24 inches (610 mm) minimum in width.

1108.4.2.1.2 Height. The top of the transfer platforms shall be 11 inches (280 mm) minimum and 18 inches (455 mm) maximum in height above the floor.

1108.4.2.1.3 Transfer Space. A transfer space complying with Sections 305.2 and 305.3 shall be provided adjacent to the transfer platform. The 48-inch (1220 mm) minimum length dimension of the transfer space shall be centered on and parallel to the 24-inch (610 mm) minimum length side of the transfer platform. The side of the transfer platform serving the transfer space shall be unobstructed.

1108.4.2.1.4 Transfer Supports. At least one means of support for transferring shall be provided.

1108.4.2.2 Transfer Steps. Transfer steps shall be provided where movement is intended from transfer platforms to levels with elevated play components required to be on accessible routes. Transfer steps shall comply with Section 1108.4.2.2.

1108.4.2.2.1 Size. Transfer steps shall have level surfaces 14 inches (355 mm) minimum in depth and 24 inches (610 mm) minimum in width.

1108.4.2.2.2 Height. Each transfer step shall be 8 inches (205 mm) maximum in height.

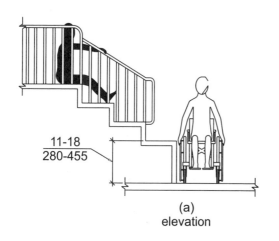

11-18
280-455

(a)
elevation

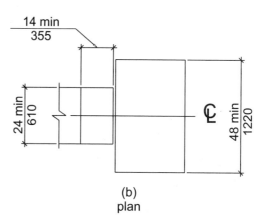

14 min
355

24 min
610

48 min
1220

(b)
plan

FIG. 1108.4.2.1
TRANSFER PLATFORMS

1108.4.2.2.3 Transfer Supports. At least one means of support for transferring shall be provided.

1108.4.3 Play Components. Ground level play components on accessible routes and elevated play components connected by ramps shall comply with Section 1108.4.3.

1108.4.3.1 Turning Space. At least one turning space complying with Section 304 shall be provided on the same level as play components. Where swings are provided, the turning space shall be located immediately adjacent to the swing.

1108.4.3.2 Clear Floor Space. Clear floor space complying with Sections 305.2 and 305.3 shall be provided at play components.

1108.4.3.3 Play Tables. Where play tables are provided, knee clearance 24 inches (610 mm) minimum in height, 17 inches (430 mm) minimum in depth, and 30 inches (760 mm) minimum in width shall be provided. The tops of rims, curbs, or other obstructions shall be 31 inches (785 mm) maximum in height.

EXCEPTION: Play tables designed and constructed primarily for children 5 years and younger shall not be required to provide knee clearance where the clear floor space required by Section 1108.4.3.2 is arranged for a parallel approach.

1108.4.3.4 Entry Points and Seats. Where play components require transfer to entry points or seats, the entry points or seats shall be 11 inches (280 mm) minimum and 24 inches (610 mm) maximum from the clear floor space.

EXCEPTION: Entry points of slides shall not be required to comply with Section 1108.4.3.4.

1108.4.3.5 Transfer Supports. Where play components require transfer to entry points or seats, at least one means of support for transferring shall be provided.

1109 Swimming Pools, Wading Pools, Hot tubs and Spas

1109.1 General. Swimming pools, wading pools, hot tubs and spas shall comply with Section 1109.

1109.1.1 Swimming pools. At least two accessible means of entry shall be provided for swimming pools. Accessible means of entry shall be swimming pool lifts complying with Section 1109.2; sloped entries complying with Section 1109.3; transfer walls complying with Section 1109.4, transfer systems complying with Section 1109.5; and pool stairs complying with Section 1109.6. At least one accessible means of entry provided shall comply with Section 1109.2 or 1109.3

EXCEPTIONS:

1. Where a swimming pool has less than 300 linear feet (91 m) of swimming pool wall, no more than one accessible means of entry shall be required.

2. Wave action pools, leisure rivers, sand bottom pools, and other pools where user access is limited to one area shall not be required to provide more than one accessible means of entry provided that the accessible means of entry is a swimming pool lift complying with Section 1109.2, a sloped entry complying with Section 1109.3, or a transfer system complying with Section 1109.5.

3. Catch pools shall not be required to provide an accessible means of entry provided that the catch pool edge is on an accessible route.

1109.1.2 Wading pools. At least one sloped entry complying with Section 1109.3 shall be provided in wading pools.

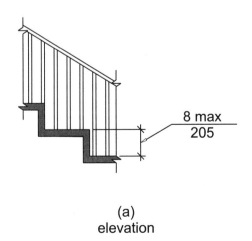
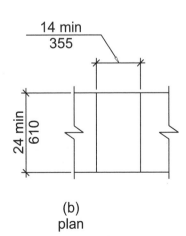

(a)
elevation

(b)
plan

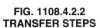

FIG. 1108.4.2.2
TRANSFER STEPS

1109.1.3 Hot tubs and Spas. At least one accessible means of entry shall be provided for hot tubs and spas. Accessible means of entry shall comply with swimming pool lifts complying with Section 1109.2; transfer walls complying with Section 1109.4; or transfer systems complying with Section 1109.5.

EXCEPTION: Where hot tubs or spas are provided in a cluster, no more than 5 percent, but not less than one hot tub or spa in each cluster shall be required to comply with Section 1109.1.3.

1109.2 Pool Lifts. Pool lifts shall comply with Section 1109.2.

1109.2.1 Pool Lift Location. Pool lifts shall be located where the water level does not exceed 48 inches (1220 mm).

EXCEPTIONS:

1. Where the entire pool depth is greater than 48 inches (1220 mm), compliance with Section 1109.2.1 shall not be required.

2. Where multiple pool lift locations are provided, no more than one pool lift shall be required to be located in an area where the water level is 48 inches (1220 mm) maximum.

1109.2.2 Seat Location. In the raised position, the centerline of the seat shall be located over the deck and 16 inches (405 mm) minimum from the edge of the pool. The deck surface between the centerline of the seat and the pool edge shall have a slope not steeper than 1:48.

1109.2.3 Clear Deck Space. On the side of the seat opposite the water, a clear deck space shall be provided parallel with the seat. The space shall be 36 inches (915 mm) minimum in width and shall extend forward 48 inches (1220 mm) minimum from a line located 12 inches (305 mm) behind the rear edge of the seat. The clear deck space shall have a slope not steeper than 1:48.

1109.2.4 Seat Height. The height of the lift seat shall be designed to allow a stop at 16 inches (405 mm) minimum and 19 inches (485 mm) maximum measured from the deck to the top of the seat surface when in the raised (load) position.

1109.2.5 Seat. The seat shall be 16 inches (405 mm) minimum in width, provide a back rest, and be of a firm and stable design.

1109.2.6 Footrests and Armrests. Footrests shall be provided and shall move with the seat. If provided, the armrest positioned opposite the water shall be removable or shall fold clear of the seat when the seat is in the raised (load) position.

EXCEPTION: Footrests shall not be required on pool lifts provided in spas.

1109.2.7 Operation. The lift shall be capable of unassisted operation from both the deck and water levels. Controls and operating mechanisms shall be unobstructed when the lift is in use and shall comply with Section 309.4.

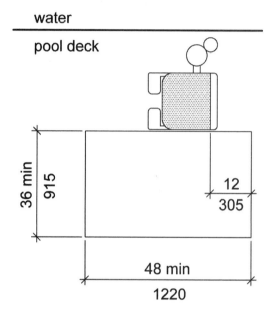

FIGURE 1109.2.3
CLEAR DECK SPACE AT POOL LIFTS

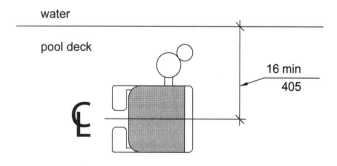

FIGURE 1109.2.2
POOL LIFT SEAT LOCATION

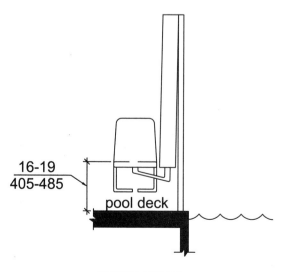

FIGURE 1109.2.4
POOL LIFT SEAT HEIGHT

1109.2.8 Submerged Depth. The lift shall be designed so that the seat will submerge to a water depth of 18 inches (455 mm) minimum below the stationary water level.

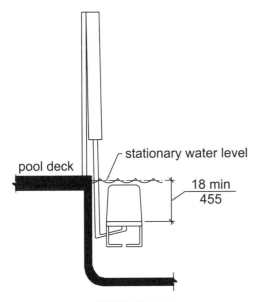

FIGURE 1109.2.8
POOL LIFT SUBMERGED DEPTH

1109.2.9 Lifting Capacity. Single person pool lifts shall have a weight capacity of 300 pounds (136 kg) minimum and be capable of sustaining a static load of at least one and a half times the rated load.

1109.3 Sloped Entries. Sloped entries shall comply with Section 1109.3.

1109.3.1 Sloped Entry Route. Sloped entries shall comply with Chapter 4 except as modified by Sections 1109.3.1 through 1109.3.3.

EXCEPTION: Where sloped entries are provided, the surfaces shall not be required to be slip resistant.

1109.3.2 Submerged Depth. Sloped entries for swimming pools shall comply with Section 1109.3.2.1. Sloped entries for wading pools shall comply with Section 1109.3.2.2.

1109.3.2.1 Swimming Pools. Sloped entries for swimming pools shall extend to a depth of 24 inches (610 mm) minimum and 30 inches (760

mm) maximum below the stationary water level. Where landings are required by Section 405.7, at least one landing shall be located 24 inches (610 mm) minimum and 30 inches (760 mm) maximum below the stationary water level.

1109.3.2.2 Wading Pools. In wading pools, the sloped entry shall extend to the deepest part of the wading pool.

1109.3.3 Handrails. At least two handrails complying with Section 505 shall be provided on the sloped entry. The clear width between required handrails shall be 33 inches (840 mm) minimum and 38 inches (965 mm) maximum.

EXCEPTIONS:

1. Handrail extensions specified by Section 505.10.1 shall not be required at the bottom landing serving a sloped entry.

2. Where a sloped entry is provided for wave action pools, leisure rivers, sand bottom pools, and other pools where user access is limited to one area, the handrails shall not be required to comply with the clear width requirements of Section 1109.3.3.

3. Sloped entries in wading pools shall not be required to provide handrails complying with Section 1109.3.3. If provided, handrails on sloped entries in wading pools shall not be required to comply with Section 505.

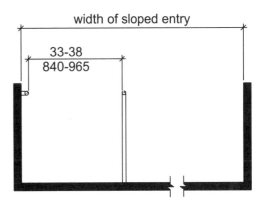

FIGURE 1109.3.3
HANDRAILS FOR SLOPED ENTRY

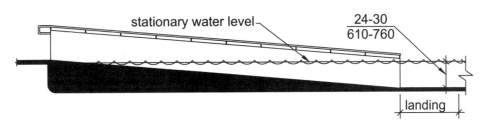

FIGURE 1109.3.2
SLOPED ENTRY SUBMERGED DEPTH

1109.4 Transfer Walls. Transfer walls shall comply with Section 1109.4.

1109.4.1 Clear Deck Space. A clear deck space of 60 inches (1525 mm) minimum by 60 inches (1525 mm) minimum with a slope not steeper than 1:48 shall be provided at the base of the transfer wall. Where one grab bar is provided, the clear deck space shall be centered on the grab bar. Where two grab bars are provided, the clear deck space shall be centered on the clearance between the grab bars.

1109.4.2 Height. The height of the transfer wall shall be 16 inches (405 mm) minimum and 19 inches (485 mm) maximum measured from the deck.

1109.4.3 Wall Depth and Length. The transfer wall shall be 12 inches (305 mm) minimum and 16 inches (405 mm) maximum in depth. The transfer wall shall be 60 inches (1525 mm) minimum in length and shall be centered on the clear deck space.

1109.4.4 Surface. Surfaces of transfer walls shall not be sharp and shall have rounded edges.

1109.4.5 Grab Bars. At least one grab bar complying with Sections 609.1 through 609.3 and 609.5 through 609.8 shall be provided on the transfer wall. Grab bars shall be perpendicular to the pool wall and shall extend the full depth of the transfer wall. The top of the gripping surface shall be 4 inches (100 mm) minimum and 6 inches (150 mm) maximum above the transfer wall. Where one grab bar is provided, clearance shall be 24 inches (610 mm) minimum on both sides of the grab bar. Where two grab bars are provided, clearance between grab bars shall be 24 inches (610 mm) minimum.

1109.5 Transfer Systems. Transfer systems shall comply with Section 1109.5.

1109.5.1 Transfer Platform. A transfer platform shall be provided at the head of each transfer system. Transfer platforms shall provide a clear depth of 19 inches (485 mm) minimum and a clear width of 24 inches (610 mm) minimum.

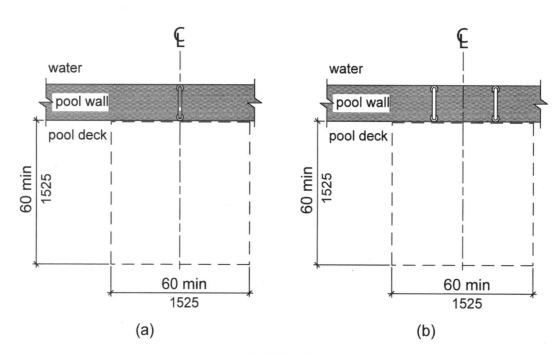

(a)   (b)

FIGURE 1109.4.1
CLEAR DECK SPACE AT TRANSFER WALLS

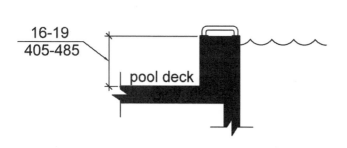

FIGURE 1109.4.2
TRANSFER WALL HEIGHT

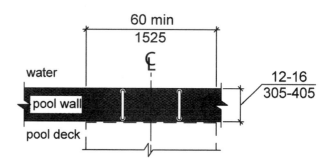

FIGURE 1109.4.3
DEPTH AND LENGTH OF TRANSFER WALLS

1109.5.2 Transfer Space. A transfer space of 60 inches (1525 mm) minimum by 60 inches (1525 mm) minimum with a slope not steeper than 1:48 shall be provided at the base of the transfer platform surface. The transfer space shall be centered along a 24-inch (610 mm) minimum side of the transfer platform. The side of the transfer platform serving the transfer space shall be unobstructed.

1109.5.3 Height. The height of the transfer platform shall comply with Section 1109.4.2.

1109.5.4 Transfer Steps. Transfer steps shall be 8 inches (205 mm) maximum in height. The surface of the bottom tread shall extend to a water depth of 18 inches (455 mm) minimum below the stationary water level.

1109.5.5 Surface. The surface of the transfer system shall not be sharp and shall have rounded edges.

1109.5.6 Size. Each transfer step shall have a tread clear depth of 14 inches (355 mm) minimum and 17 inches (430 mm) maximum and shall have a tread clear width of 24 inches (610 mm) minimum.

1109.5.7 Grab Bars. At least one grab bar on each transfer step and the transfer platform or a continuous grab bar serving each transfer step and the transfer platform shall be provided. Where a grab bar is provided on each step, the tops of gripping surfaces shall be 4 inches (100 mm) minimum and 6 inches (150 mm) maximum above each step and transfer platform. Where a continuous grab bar is provided, the top of the gripping surface shall be 4

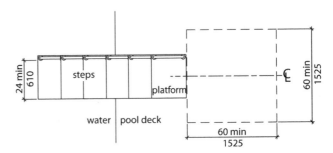

FIG. 1109.5.2
CLEAR DECK SPACE AT TRANSFER PLATFORM

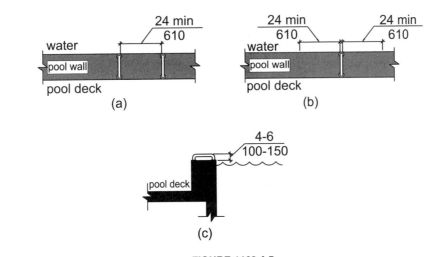

FIGURE 1109.4.5
GRAB BARS FOR TRANSFER WALLS

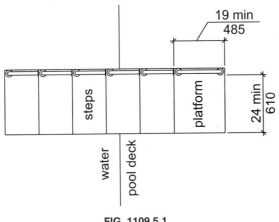

FIG. 1109.5.1
SIZE OF TRANSFER PLATFORM

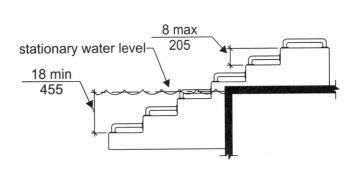

FIG. 1109.5.4
TRANSFER STEPS

inches (100 mm) minimum and 6 inches (150 mm) maximum above the step nosing and transfer platform. Grab bars shall comply with Sections 609.1 through 609.3 and 609.5 through 609.8 and be located on at least one side of the transfer system. The grab bar located at the transfer platform shall not obstruct transfer.

1109.6 Pool Stairs. Pool stairs shall comply with Section 1109.6.

1109.6.1 Pool Stairs. Pool stairs shall comply with Section 504.

EXCEPTION: Pool step risers shall not be required to be 4 inches (100 mm) minimum and 7 inches (180 mm) maximum in height provided that riser heights are uniform.

1109.6.2 Handrails. The width between handrails shall be 20 inches (510 mm) minimum and 24 inches (610 mm) maximum. Handrail extensions required by 505.10.3 shall not be required on pool stairs.

1110 Shooting Facilities with Firing Positions

1110.1 Turning Space. A circular turning space complying with Section 304.3.1 with slopes not steeper than 1:48 shall be provided at shooting facility firing positions.

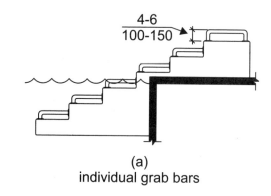

(a)
individual grab bars

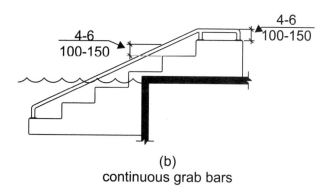

(b)
continuous grab bars

FIG. 1109.5.7
GRAB BARS

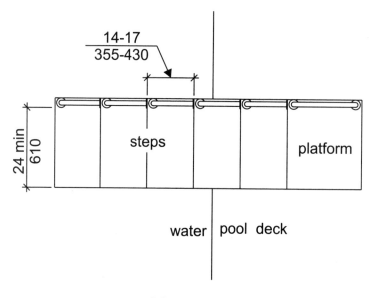

FIG. 1109.5.6
SIZE OF TRANSFER STEPS

Index

Word	See also	References
A		
Access Aisles	Parking, Passenger loading zone	502.4, 503.3
Accessible		106.5
Accessible Routes		Chapter 4, 402, 502.8, 1002.3, 1003.3, 1004.3, 1101.2.1, 1102.2, 1103.2, 1105.1, 1106.2, 1107.2,1108.2, 1108.4
Accessible unit		101, 202, 1001, 1002
Accessible route		1002.3
Beds		1002.15
Doors & doorways		1002.5
Elevator		1002.7
Kitchen		1002.12
Laundry equipment		1002.10
Operable part		1002.9
Platform lift		1002.8
Primary entrance		1002.2
Ramps		1002.6
Reinforcement		1002.11, 1002.11.1
Storage		1002.14
Toilet and bathing facility		1002.11, 1002.11.2
Walking surface		1002.4
Window		1002.13
Addition		1108.3.1
Administrative authority		103, 106.5, 201, 202, 203, 301.1, 401.1, 404.2.8, 501.1, 601.1, 701.1, 801.1, 807.3, 901.1, 1001.1, 1101.1
Airports		805.1
Clock		805.8
Escalator		805.9
Public Address System		805.7
Variable Message Signage		703.7
Aisle Seats	Designated Aisle Seats	
Aisles	Parking, Passenger Loading zones, Assembly areas, Checkout aisles	
Alarms		105.2.2, 702, 806.3.1, 1006.2, 1006.3, 1006.4
Alcove		305.7
Alteration	Existing facilities	201, 308.3.1, 404.2.2, 404.2.4, 406.7, 505.10, 807.3, 1103.2, 1108.3

(continued)

Word	See also	References
Alternative		103
Ambulatory stall	Toilet compartments	604.10
American National Standards Institute-ANSI		Forward, 105.2
Amusement Ride		106.5, 1101.2.1, 1102
Animal Containment Area		1101.2.1
ANSI–American National Standards Institute		Forward, 105.2
Anthropometrics		102
Appliance	Accessible unit, Kitchen, Type A units, Type B units, Type C units	804.5, 1002.12, 1003.12.5, 1004.12.2, 1005.7
Alarm notification	Alarms	702.1, 1006.4
Clothes dryer		611, 1002.10, 1003.10, 1004.10
Controls	Operable parts	1002.9, 1003.9, 1004.9
Cooktop		804.5.4, 1003.12.5.4, 1004.12.2.3
Dishwasher		804.5.3, 1003.12.5.3, 1004.12.2.2
Oven		804.5.5, 1003.12.5.5, 1004.12.2.4
Range	Cooktop, oven	
Refrigerator/freezer		804.5.6, 1003.12.5.6, 1004.12.2.5, 1005.7
Trash compactor		1004.12.2.6
Washing machine		611, 1002.10, 1003.10, 1004.10
Area of Sport Activity	Sports activity	106.5, 1101.2.2, 1101.3
ASME- American Society of Mechanical Engineers, International	CSA	105.2.5, 105.2.6, 407.1, 407.4.3, 407.4.5, 407.4.10, 408.1, 408.4.3, 409.1, 410.1, 805.9
Assembly Area	Seats, assembly	802
Assistive Listening System		706
ASTM–ASTM International		105.2.8, 105.2.9, 105.2.10, 106.5 – 'use zone', 1108.4.1.6.2
ATM	Automatic Teller Machines	707
Audible Alarm	Alarms	702.1, 806.3.1, 1006.2
Audible sign	Sign	703.8
Automatic Door	Doors and Doorways	105.2.3, 105.2.4, 404.3, 407.3.2, 408.3.2, 1004.5.2.3
Automatic Fare Machine		707
Automatic Teller Machine		707
B		
Bathing Room/Bathroom	Water closet, Lavatory, Shower , Bathtubs	603, 806.2.4
Children's use	Children	
Dwelling and sleeping unit		1002.11, 1002.11.2, 1003.11, 1003.11.2, 1004.11, 1004.11.3, 1005.4, 1005.6

(continued)

Word	See also	References
Private office, accessed through a		603.2.3, 604.4, 604.5, 606.2, 606.3, 607.4, 608.2.1.3, 608.2.2.3, 608.2.3.2, 608.3
Single occupant		603.2.2, 604.4, 604.5, 606.2, 606.3, 607.4, 608.2.1.3, 608.2.2.3, 608.2.3.2, 608.3
Bathtub		607, 1002.11.2, 1003.11.2, 1003.11.2.5, 1004.11.3.1.3, 1004.11.3.2.3
Bathtub Seat	Seat, tub and shower	610.2
Bed		806.2.3, 1002.15
Bench	Children	612.2, 803.4, 806.2.2, 807.4, 903
BHMA–Builder's Hardware Manufacturer's Association		105.2.3, 105.2.4, 404.3, 408.3.2.1, 409.3.1
Boarding Pier	Boating facility	106.5, 1103.2.2, 1103.3.2
Boat Launch Ramp		106.5, 1103.2.2, 1103.3.2
Boat Slip	Boating facility	106.5, 1103.2.1, 1103.3
Boating Facility		1101.2.3, 1103
Bowling Lane		1101.2.1
Boxing Ring		1101.2.1
Braille		407.4.7.1.2, 407.4.7.1.3, 504.9, 703.3.1, 703.4, 707.10
Building		101, 201
Bus Shelter		805.3
Bus Station		805.1
Boarding and Alighting area		805.2
Clock		805.8
Escalator		805.9
Public Address System		805.7
Shelters		805.3
Sign		805.4
Variable Message Signage		703.7
Bus Stop		805.2, 805.3, 805.4
C		
Cafeteria	Dining surfaces, tray slides, self-service food line	
Carpet	Floor or floor surface	302.2
Catch Pool	Swimming Pool	106.5, 1109.1.1
Cell, Jail/Holding		604.5, 806, 904.6
Changes in Level	Curb Ramp, Ramp, Stairway	303
Character	Signs	106.5
Raised		703.3
Visual		703.2

(continued)

Word	See also	References
VMS		106.5
Check-Out Aisle		904.4
Chair lift	Platform lift	105.2.6, 410, 1002.8, 1003.8, 1004.8
Children's		102, 106.5
Amusement rides		1101.2.1
Bench		903.5
Dining surface		902.1, 902.5
Drinking fountain		602.2
Lavatory		606.2
Play area		1108.3, 1108.4.3.3
Sink		606.2
Toilet compartment		604.1, 604.9.2.2, 604.9.5, 604.11, 609.4.2
Water closet		604.1, 604.11, 609.4.2
Work surface		902.1, 902.5
Clear Floor Space		301.2, 305, 309.2
Clearance	Doors and Doorways, knee clearance, toe clearance	
Clock		805.8
Clothes Dryer		611, 1002.10, 1003.10, 1004.10
Coat Hook		603.4, 604.3.3, 604.8, 803.5, 1003.11.2.4.4
Common Use	Access aisle, Bathing room, Communication system, Toilet room	
Communication System	TTY, Two-Way Communication Systems	105.2.7, 202, 407.4.10, 408.4.8, 409.4.7, 701.1, 708, 806.3, 904.6, 1006.5, 1006.6, 1006.7
Commuter Rail	Rail Station	
Companion Seat	Wheelchair space location	802.7, 1102.4.7
Compliance Alternatives		103
Convention		104
Cooktop	Appliance	804.5.4, 1003.12.5.4, 1004.12.2.3
Correctional Facility	Cells, Security glazing	
Counter	Dining Surface, Work Surface, Lavatory, Tray Slide, Self-service food line, Sales counter, Service counter, Checkout aisles	603.3, 1002.11.2.1
Courtroom		807
CSA-Canadian Standards Association	ASME	105.2.5, 407.1, 407.4.3, 407.4.5, 407.4.10, 408.4.3, 409.1, 805.9
Curb Cut	Curb Ramp	
Curb Ramp	Ramp	106.5, 402.2, 406

(continued)

Word	See also	References
D		
DASMA – Door and Access Systems Manufacturers Association		105.2.7, 708.4
Defined Terms	Definition	106.5
Definition	Defined Terms	106
Designated Aisle Seat	Seats, assembly	802.8
Destination Oriented Elevator	Elevators	106.5, 407
Detectable Warning		106.5, 406.12, 406.13, 406.14, 705, 805.5.2, 805.10
Detention and Correctional Facility	Cells	
Diaper changing tables		603.5
Dimension		104.2
Dining Surface	Children's	902
Dishwasher	Appliance	
Dispersion	Seat-assembly, Line of sight, Integration	802.10, 1105.2.1.1, 1108.3.2.1
Diving Boards and Diving Platform	Catch pools	1101.2.1
Door Swing	Door and Doorway – swinging	304.4, 404.3.5, 603.2.2, 604.9.3, 604.10.3, 612.2, 703.3.11, 803.3, 1003.11.2.1, 1004.11.2.1
Door and Doorway		301.2, 402.2, 404
Automatic		404.3, 407.3.2, 407.3.3. 408.3.2
Clearance (width)		404.2.2, 404.3.1, 407.3.6, 408.3.3, 409.3.3, 410.2.1, 1004.5.2.1, 1005.5.3.1
Dressing, Fitting and Locker Room		803.3
Dwelling and Sleeping unit		1002.5, 1003.5, 1004.5, 1005.5.3
Maneuvering Clearance		404.2.3, 404.3.2, 409.3, 604.9.3, 604.10.3, 1003.5,
Manual		404.2
Revolving		402.3
Swinging	Door Swing	404.2.2, 404.2.3.2, 404.2.5, 407.3.2, 408.3.2, 409.3, 1004.5.2.1, 1005.5.3.1
Toilet and Bathing Room		603.2.2, 1003.11.2.1, 1004.11.2.1
Toilet Stall		604.9.3, 604.10.3
Dressing Room		301.2, 803
Drinking Fountain		602
Children's use		602.2
Standing person		602
Wheelchair accessible		602
Dryer	Appliance–clothes dryer, Hand dryer	

(continued)

Word	See also	References
Dwelling unit	Accessible unit, Sleeping unit, Type A unit, Type B unit, Type C unit, Communication systems, Alarms	106.5, 202, Chapter 10
E		
Edge Protection		405.9, 904.4.2, 1005.5.4, 1103.3.1, 1103.3.2, 1105.3
Electrical panelboards		1002.9, 1003.9, 1004.9
Element		101.1, 106.5, 201, 501.1, 601.1
Elevator		105.2.5, 106.5, 402.2, 407, 1002.7, 1003.7, 1004.7
Destination Oriented		106.5, 407
Existing		407
LULA–Limited Use / Limited Application		408
Private Residence		409
Entrance		805.6.1, 1002.2, 1002.5, 1003.2, 1003.5, 1004.2, 1004.5.1, 1005.2, 1006.5, 1006.6
Exercise Machine		1101.2.4, 1104
Existing Facilities		201, 308.3.1, 404.2.2, 404.2.4, 405.2, 406.7, 407, 408.4.1, 505.10, 608.7, 805.5.1, 805.9, 807.3, 1002.5, 1103.2.1, 1108.3
F		
Facility		101.1, 106.5, 201
Fare Machine		707
Figure		104.3
Fire Alarm System		105.2.2, 702, 1006.3, 1006.4
Fishing Piers and Platform		1105
Fitting room		803
Floor or Floor Surface	Walking Surface	104.4, 302, 303, 304.2, 305.2, 403.2, 404.2.3.1, 405.4, 405.9.1, 406.5, 407.4.2, 408.4.2, 409.4.2, 410.3, 502.5, 503.4, 802.2, 1002.4, 1003.4, 1004.4, 1005.5.2, 1102.4.1, 1102.4.2
Food Service Line		904.5
Forward Reach	Reach Ranges	
Freezer	Appliance	
G		
Gate	Doors and Doorways	
Golf Car Passage	Golf facility	106.5, 1106.3
Golf Facility	Miniature golf facility	106.5, 1106

(continued)

Word	See also	References
Grab Bars	Reinforcement	609, 1002.11.1, 1003.11.1, 1004.11.1, 1005.6, 1109.4.5, 1109.5.7
Ambulatory stall		604.10.4
Bathtub		607.4
Children's water closet		604.11.5, 609.4.2
Pools		1109.4.1, 1109.4.5, 1109.5.7
Shower		608.3
Water closet		604.5, 604.11.5
Wheelchair stall		604.9.6
Grating		302.3
Ground Level Play Component	Play Area	106.5, 1108.2.1, 1108.3.2.1, 1108.4.3
Ground Surface	Floor or Floor Surface, Use Zone	104.4
H		
Handrail		403.6, 405.8, 406.9, 504.6, 505, 1105.2, 1108.4.1.5.3, 1109.3.3, 1109.6.2
High Speed Rail	Rail station	
Hoistway	Elevator	Elevator
Holding Cell	Cell	
Hospital		1002.5
Hot tub	Pools	
Housing Cell	Cell	
I		
ICC–International Code Council		Forward
Integration	Seat–assembly, Dispersion, Line of sight	802.6, 1108.3.2.1
Intercity Rail	Rail station	
J		
Jail	Cells	806
Judges' Benches	Courtroom	807.2, 807.4
Judge's Stand-sports		1101.2.1
Judicial Facility	Cell, Courtroom	806, 807
Jury Box		807.2, 807.3
K		
Kitchen and Kitchenette		301.2, 606, 804, 1002.12, 1002.14, 1003.12, 1003.14, 1004.12, 1005.7

(continued)

Word	See also	References
Knee Clearance	Toe Clearance	301.2, 304.3.1, 304.3.2, 305.4, 306.1, 306.3, 404.2.3, 602.2, 606.2, 608.2.2.2, 804.3, 804.4, 804.5.4.2, 902.2, 902.5.1, 904.3.2, 1003.11.2.2, 1003.12.3.1, 1003.12.4.1, 1003.12.5.4.2, 1004.11.2.1, 1004.11.2.2, 1004.11.3.1.1, 1004.11.3.1.3.1, 1004.12.2.3.2, 1108.4.3.3
L		
Laundry Equipment	Appliance, Clothes Dryer, Washing Machine	
Lavatory	Sinks, Children	301.2, 606, 1002.11.2, 1003.11.2, 1003.11.2.2, 1004.11.3.1, 1004.11.3.1.1, 1004.11.3.2.1, 1005.6
Lift	Platform Lift, Elevator	
Light Rail	Rail station	
Limited-Use/Limited-Application Elevator	Elevator	408
Lines of Sight	Dispersion, Integration, Seats-assembly	802.9
Locker	Storage	905
Locker Room		803, 903
LULA Elevator	Elevator	408
M		
Maneuvering Clearance	Doors & Doorways	305.7, 404.2.3, 405.7.5, 1102.4.4.2
Manual door	Doors & Doorways	404.2
Mezzanine		805.6.2, 1004.3.1
Miniature Golf Facility		1107
Mirror		603.3, 1002.11.2.2, 1003.11.2.3
Motion picture projection viewing		802.10, 802.10.2, 802.10.4
MUTCD–Manual on Uniform Traffic Control Devices		105.2.1, 703.9
N		
NFPA–National Fire Protection Association		105.2.2, 702, 1006.2, 1006.3, 1006.4
O		
Operable Part		106.5, 309, 404.2.6, 407.2.1.1, 407.4.10.1, 409.4.7.2, 410.6, 506.1, 602.3, 603.6, 611.3, 704.2.2, 707.3, 804.5.2, 905.4, 1002.9, 1002.13, 1003.9, 1003.12.5.1, 1003.13, 1004.5.2.1.1, 1004.9, 1005.8, 1006.6, 1101.2.3

(continued)

Word	See also	References
Oven	Appliance	
Overlap		301.2, 304.4, 405.7.5, 406.11, 502.4.1, 503.3.1, 603.2.2, 604.3.3, 604.9.3, 604.10.3, 612.2, 802.5, 803.3, 1003.11.2.1, 1003.11.2.4.4, 1004.11.2.1, 1004.11.3.1.2.2.4, 1102.4.4, 1102.5.4, 1102.6.3, 1104.1

P

Word	See also	References
Parking		406.6, 406.8, 502
Access aisle		502.4
Floor surface		502.5
Identification		502.7
Spaces		502
Vehicle Space Marking		502.3
Vehicle Space Size		502.2
Vertical Clearance		502.6
Passenger Loading Zone		503
Passing Space		403.5.2
Pictogram	Signs	106.5, 703.1.3, 703.5, 703.6.3.4
Platform Lift		105.2.6, 402.2, 410, 807.3, 1002.8, 1003.8, 1004.8, 1005.5.1
Play Area		105.2.8, 105.2.9, 105.2.10, 106.5, 1101.3, 1108
Plumbing Fixture	Drinking fountain, Sink, Lavatory, Water closet, Bathtub, Shower, Kitchen & Kitchenette	
Pool	Catch pool	106.5, 1101.2.1, 1109
Pool lift	Pool	1109.2
Pool stairs	Pool	1109.6
Post-Mounted Object	Protruding object	
Power door	Doors & Doorways-automatic	
Private Residence Elevator	Elevator	409
Protruding Object		307, 407.2.1, 704.7, 1101.3, 1102.4.4.3
Public Address System		805.7
Public Entrance	Entrances	
Putting Green	Golf facilities	1106.2

R

Word	See also	References
Rail Station		805.1
Clock		805.8
Escalator		805.9
Platform		805.5

(continued)

Word	See also	References
Public Address System		805.7
Sign		805.6
Track crossing		805.10
Variable Message Signage		703.7
Raised Area	Mezzanines	1104.3.1
Raised Character	Signs	
Ramp	Curb Ramp	106.5, 303.4, 402.2, 405, 1002.6, 1003.6, 1004.6, 1005.5.4, 1102.2, 1103.2.1, 1107.2, 1108.4.1.5
Range	Appliance	
Reach Ranges		308, 309.3, 1107.3.2
Enhanced reach		603.6, 606.5
Recreation Facilities	Amusement Ride, Boating Facility, Exercise Machine, Fishing Pier and Platform, Golf, Miniature Golf, Sauna and Steam Room, Shooting Facility, Swimming Pool, Wading Pool, or Spa	Chapter 11
Recreational Boating Facility	Boating Facility	
Referee's stand		1101.2.1
Referenced Standard		105
Refrigerator	appliances	
Reinforcement		1002.11.1, 1003.11.1, 1004.11.1, 1005.6
Residential Dwelling Unit	Accessible unit, Dwelling unit, Sleeping unit, Type A unit, Type B unit, Type C unit	106.5, Chapter 10
Restaurant	Dining surface	
Rug	Carpet	
S		
Sales Counter		904.3
Sauna		612
Scoring stand		1101.2.1
Seat, Assembly		802
Companion seat		802.7
Courtroom gallery		807.5
Designated aisle seat		802.8
Dispersion		802.10
Integration		802.6
Line of sight		802.9
Motion picture projection		802.10, 802.10.2, 802.10.4
Wheelchair space		106.5, 802.1, 802.3, 802.4, 802.5, 802.7, 807.3

(continued)

Word	See also	References
Wheelchair space location		106.5, 802.1, 802.2, 802.6, 802.9, 802.10
Seat, Tub and Shower		607.3, 608.2.1.3, 608.2.2.3, 608.2.3.2, 610
Security Glazing		904.6
Self-Service Food Line		904.5
Service Counter		904.3
Shelf		603.4, 604.8, 704.6, 803.5, 804.5.6, 904.5.1, 1003.12.5.6
Shooting Facility		1101.1, 1110
Shower Compartments		608, 1002.11.2, 1003.11.2, 1003.11.2.5.2, 1004.11.3.1.3.3, 1004.11.3.2.3.2
Alternate Roll-in		608.2.3
Standard Roll-in		608.2.2
Transfer		608.2.1
Shower Seat	Seats, tub & shower	
Sign		106.5, 703
Braille		703.4
Bus sign		805.4
Elevator destination		407.2.4
Hoistway		407.2.3, 408.2.3
Parking		502.7
Pictogram		106.5, 703.1.3, 703.5, 703.6.3.4
Rail Station		805.6
Raised character		703.3
Remote infrared audible sign		703.8
Stairway identification		504.9
Symbol		703.6
Variable message		106.5, 703.7
Visual character		703.2
Sink	Lavatory, Children	606, 804.4, 1003.12.4, 1003.12.5.3, 1004.12.2.1, 1005.7
Site		106.5, 201
Sleeping unit	Accessible unit, Dwelling unit, Type A unit, Type B unit, Communication systems, Alarms	106.5, 202, Chapter 10
Slope	Ramp, Curb ramp	
Sloped entry	Pool	1109.3
Soft Contained Play Structure	Play Area	106.5, 1108.2.2, 1108.4.1.2
Smoke detector		1006.2, 1006.3, 1006.4
Spa	Pool	

(continued)

Word	See also	References
Sports activity		106.5, 1101.2.1, 1101.2.2, 1101.3
Stairway		504
Standard	Referenced standard	
Steam Room		612
Storage		905, 1002.14, 1003.14, 1102.5.4, 1102.6.3
Swimming Pool	Pool	
Symbol	Sign	703.6
T		
Tactile	Signs-braille, Signs-raised character	703.1
TDD	TTY	
Teeing Ground	Golf facility	106.5, 1106.2
Telephone		409.4.7, 703.6.3.4, 704, 708.4, 806.3.2, 1006.6.2
Temporary Facility		201
Threshold		303, 404.2.4, 404.3.3, 608.6, 1003.5, 1004.5.2.2, 1005.5.3.2
Toe Clearance	Knee Clearance	301.2, 304.3.1, 304.3.2, 305.4, 306.1, 306.2, 404.2.3, 602.2, 604.9.5, 606.2, 704.2.1.2, 804.3, 804.5.4.2, 902.2, 902.5.1, 904.3.2, 1003.12.3.1, 1003.12.4.1, 1003.12.5.4.2, 1004.11.2.1, 1004.11.2.2, 1004.11.3.1.1, 1004.11.3.2.1, 1004.12.2.1, 1004.12.2.3.2, 1105.3.2,
Toilet	Water closet	
Toilet Compartment		604
Ambulatory		604.10
Children's use	Children	604.9.2.2, 604.11
Wheelchair		604.9
Toilet Room	Water closet, Lavatory	301.2, 603, 806.2.4
Children's use	Children	
Dwelling and sleeping unit		1002.11, 1003.11, 1004.11, 1005.4, 1005.6
Private office, accessed through a		603.2.2, 604.4, 604.5, 606.2, 606.3, 607.4
Single occupant		603.2.2, 604.4, 604.5, 606.2, 606.3, 607.4
Toilet Stall	Toilet Compartment	
Tolerances, Construction and Manufacturing		104.2
Train Station	Rail Station	
Transfer System	Pool	1108.4.2, 1109.1.1, 1109.1.3, 1109.5

(continued)

Word	See also	References
Transfer wall	Pool	1109.1.1, 1109.4
Transportation Facility	Airport, Bus Station, Bus Stop, Rail Station	705.6, 805
Tray Slide		904.5
Truncated Dome	Detectable Warning	
TTY		106.5, 704.4, 704.5, 704.6, 704.7
Turning Space		301.2, 304, 404.2.5, 405.7.4, 405.7.5, 603.2.1, 612.3, 803.2, 806.2.1, 807.2, 1002.3.2, 1002.5, 1003.3.2, 1003.5, 1102.3, 1105.5, 1108.4.1.3, 1108.4.1.4.1, 1108.4.1.6, 1108.4.3.1, 1110.1
Two-Way Communication System	Communication systems	708
Type A unit		101, 1001, 1003
Accessible route		1003.3
Doors & doorways		1003.5
Elevator		1003.7
Kitchen and kitchenette		1003.12
Laundry equipment		1003.10
Operable part		1003.9
Platform lift		1003.8
Primary entrance		1003.2
Ramp		1003.6
Reinforcement		1003.11, 1003.11.1
Storage		1003.14
Toilet and bathing facility		1003.11
Walking surface		1003.4
Window		1003.13
Type B unit		101, 1001, 1004
Accessible route		1004.3
Doors & doorways		1004.5
Elevator		1004.7
Kitchen and kitchenette		1004.12
Laundry equipment		1004.10
Operable part		1004.9
Platform lift		1004.8
Primary entrance		1004.2
Ramp		1004.6
Reinforcement		1004.11, 1004.11.1
Toilet and bathing facility		1004.11
Walking surface		1004.4

(continued)

Word	See also	References
Type C (Visitable) units		101, 1001, 1005
Circulation path		1005.5
Connected spaces		1005.3
Food preparation area		1005.7
Interior space		1005.4
Lighting controls and receptacle outlets		1005.8
Toilet room or bathroom		1005.6
Unit entrance		1005.2
U		
Unisex Toilet Room	Toilet room	
Use Zone	Play area	105.2.9, 105.2.10, 106.5, 1108.3, 1108.4.1.5.3, 1108.4.1.6.2
Urinals		605
V		
Van Parking Space	Parking	502
Variable Message Sign	Sign	106.5, 703.7
High-Resolution		106.5 – VMS characters, 703.7.1
Low-Resolution		106.5 – VMS characters, 703.7.1
Vertical Clearance	Protruding objects, Parking, Passenger loading zone	307.4, 502.6, 503.5, 704.6, 1101.3, 1108.4.1
Vision Lite	Doors and Doorways	404.2.10
Visual Alarm	Alarm	
W		
Wading Pool	Pool	
Walk		106.5, 406.2
Walking Surface	Floor or floor surface	402.2, 403, 1002.4, 1003.4, 1004.4, 1005.5.2
Washing Machine	Appliance	611, 1002.10, 1003.10, 1004.10
Water Closet	Children	604.1 to 604.7, 604.11, 609.4.2, 1003.11.2, 1003.11.2.4, 1004.11.3.1.2, 1004.11.3.2.2, 1005.6
Water Slide	Catch pool	1101.2.1
Weather Shelters	Golf Facilities	1106.4
Wheelchair Space	Seat, Assembly	106.5, 802.1, 802.3, 802.4, 802.5, 802.7, 807.3, 1101.2.1, 1102.4
Wheelchair space location	Seat, assembly	106.5, 802.1, 802.2, 802.6, 802.9, 802.10
Window		506, 1002.13, 1003.13
Work Surface	Children	804.3, 804.5.5.2, 804.5.5.3, 807.4, 902, 1002.12, 1003.12.3
Wrestling ring		1101.2.1

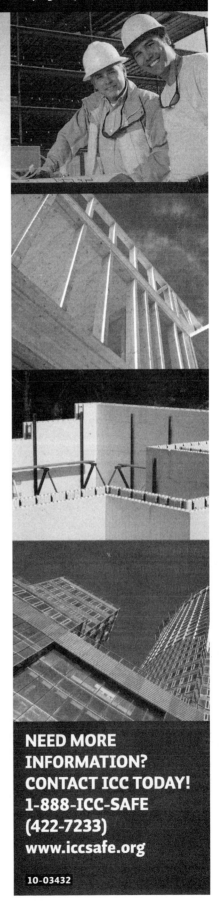

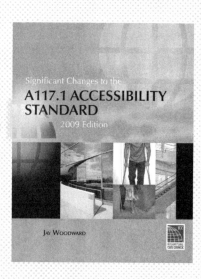